# A Pictorial Handbook
## of
# Anatomy
## and
# Physiology

### Dr James Bevan

*Introduction by Sir Richard Bayliss*
*Fellow of the Royal College of Physicians*

# Advisory Panel

Dr J. H. Baron, DM, FRCP
Consultant Physician, St Mary's Hospital, London
Senior Lecturer and Consultant, Royal Postgraduate
Medical School and Hammersmith Hospital, London

Professor G. M. Besser, MD, FRCP
Professor of Endocrinology, St Bartholomew's Hospital, London

Mr Michael Cameron, FRCS, FRCOG
Consultant Obstetrician and Gynaecologist,
St Thomas' Hospital, London

Dr Desmond Croft, DM, FRCP
Consultant Physician and Clinical Director of Nuclear Medicine,
St Thomas' Hospital, London

Mr D. Garfield Davis, FRCS
Consultant Otorhinolaryngologist, Middlesex Hospital, London
Director of Ferens Institute

the late Dr Max Friedman, MD, PhD, FRCP
Consultant Paediatrician,
University College Hospital, London

Mark Oberg-Brown, PhD

Ms Shirley Orbell, Nurse Tutor

the late Sir Henry Osmond-Clarke, KCVO, CBE, FRCS
Honorary Consultant Orthopaedic Surgeon, The London Hospital

Sir George Pinker, KCVO, FRCS, past PRCOG
Consultant Obstetrician, St Mary's Hospital, London

Professor Neil Pride, MD, FRCP
Professor of Respiratory Medicine and Consultant Physician,
Royal Postgraduate Medical School and Hammersmith Hospital, London

Ms Phaedra Dhmis, BSc, Mem.BDA
Senior Dietitian, The Wellington Hospital, London

First published in 1978 by Mitchell Beazley,
an imprint of Reed Consumer Books Limited
Michelin House, 81 Fulham Road, London SW3 6RB
and Aukland, Melbourne, Singapore and Toronto

ISBN 1 85732 392 0

Typeset by Dorchester Typesetting Limited
Produced by Mandarin Offset
Printed in Hong Kong

Editor   Lawrence Clarke

Art Editor   John Ridgeway

Assistant Editor   Cynthia Hole

Editorial Assistant   Rosemary Mendus

Production   Julian Deeming

Indexer   Valerie Nicholson

# Contents

# Introduction

There is an ever present need for a text – not a heavy academic textbook – on human anatomy and physiology for those who are interested in the workings of the human body. Such information is required by physiotherapists early in their training, social workers, councillors, medical technicians in many different hospital departments, nursing aids, and students who are either at school or involved in postgraduate studies.

Academics are seldom the most suitable people to write such a book because they are likely to be pure anatomists or pure physiologists. Dr James Bevan studied both these major subjects in depth during his many years of preclinical training. Subsequently he has assimilated the newer knowledge and the important advances. As a practising doctor he has made it his business to keep up-to-date, and in the day-to-day application of anatomy and physiology he is in a unique position to write in a style and at a level appropriate to his readers.

Since the first successful edition, changes have occurred that call for this second edition. Human anatomy may not have changed much, but by contrast physiology has advanced greatly. The chapters on nutrition, for example, have needed major revision. The working of the cell has been enlightened by the emergence of molecular biology. Endocrinology is no longer a subject confined to the endocrine glands. Local chemical factors control the functioning of many tissues, such as the heart, vascular tissue and the intestinal system. Dr Bevan has revised a book that surely will meet the needs of many.

Sir Richard Bayliss, KCVO, MD, FRCP

# Suggested Reading

**Anatomy and Physiology**
*N* Churchill Livingstone

**Principles of Anatomy and Physiology**
*N.P* Tortora and Grabowski
Harper Collins

**Human Anatomy and Physiology**
*N.P.D.* Marieb
Benjamin Cummings

**Human Nutrition and Dietetics**
*M.N.D.* Davidson
Churchill

**Gray's Anatomy**
*M* Longman

**Lecture Notes on Human Physiology**
*M* Bray
Blackwell Scientific

**Cunningham's Textbook of Anatomy**
*M* Romanes
Oxford

**Principles of Physiology**
*M* Berne Ley
Wolfe

**Clinical Anatomy for Medical Students**
*M* Snell
Little Brown

**Oxford Concise Medical Dictionary**
*M.N.P.D*

Key
*M* – Medical Students
*N* – Nurses
*P* – Physiotherapists
*D* – Dietitian

# Medical Terms

A great many apparently difficult medical terms are constructed from comparatively few elements. Once these elements are recognized many pieces of medical jargon read quite naturally.

| element | meaning *example* definition |
|---|---|
| **a** | without *a-morphous:* without shape |
| **ab** | away from *ab-duct:* to lead away from |
| **ad** | towards *ad-renal:* towards the kidney |
| **aden, adeno** | gland *aden-itis:* inflammation of a gland |
| **aesthes** | feel *an-aesthes-ia:* without feeling |
| **alg, alge** | pain *an-alge-sia:* without pain |
| **an** | without *an-algesia* |
| **andro** | *male* andro-gen: of male origin |
| **angi, angio** | vessel *angio-gram:* a record of a blood vessel, usually made by special X-ray technique |
| **ante** | before *ante-natal:* (examination carried out) before the birth |
| **arter, arterio** | artery *arterio-sclerosis:* hardening of the arteries |
| **arthr, arthro** | joint *arthr-itis:* inflammation of joints |
| **bil, bili** | bile *bili-ary:* concerning the biliary system |
| **brachi** | arm *brachi-al:* concerning the arm |
| **brady** | slow *brady-cardia:* a condition of slowing of the heart |
| **bronch, bronchi** | windpipe *bronchi-tis:* inflammation of the windpipe |
| **carcin, carcino** | cancerous *carcino-genic:* something that causes cancer |
| **cardi, cardio** | heart *cardio-myopathy:* a sickness of the muscle of the heart |
| **cele** | swelling *cysto-cele:* swelling of a bladder |
| **cephal, cephalo** | head *a-cephal-ous:* without a head |
| **cervic** | neck *cervic-itis:* inflammation of the neck of a structure |
| **chol, chole** | bile *chol-angitis:* inflammation of the bile vessels |
| **chondr, chondri** | cartilage *chondr-oma:* swelling of cartilage |
| **cortex, cortico** | outer part *cortico-trophin:* something that causes growth of the cortex |
| **cost** | rib *cost-al margin:* the edge of the ribs |
| **crani, cranio** | skull *cranio-tomy:* cutting the skull |
| **cryo** | freezing *cryo-surgery:* surgery using a freezing technique |
| **crypt, crypto** | hidden *crypto-genic disease:* disease in which the origin is hidden |
| **cut** | skin *cut-icle:* thin skin |
| **cyst** | bladder *chole-cyst-ectomy:* the removal of the gall bladder |
| **cyt, cyto** | cell *nucleo-cyt-e:* cell with nucleus |
| **dent** | tooth *dent-ist:* |
| **derm, dermato** | skin *intra-derm-al:* within the skin |
| **dis** | undo *dis-organize:* prevention of the normal working |
| **dors, dorso** | back *dorso-lumbar:* back and lumbar region |
| **dys** | bad, disorder *dys-entery:* disorder of the intestine |
| **ectomy** | cut out *gastr-ectomy:* removal of the stomach |
| **en** | in *en-cephalitis:* inflammation of the inside of the brain |
| **end, endo** | inside *endo-scopy:* to look inside |

| element | meaning *example* definition |
|---|---|
| **enter** | intestine *gastro-enter-itis:* inflammation of the stomach and intestine |
| **erythro** | red *erythro-cyte:* red cell |
| **ex** | out of *ex-hale:* to breathe out |
| **extra** | beyond *extra-systole:* a systole beyond the normal ones |
| **form** | shape *lenti-form:* shaped like a lentil |
| **gastr, gastro** | stomach *gastr-ectomy:* removal of the stomach |
| **gen** | iginate *cyto-gen-ic:* originating within the cell |
| **gloss, glosso** | tongue *glosso-pharyngeal:* concerning the tongue and pharynx |
| **gram** | recording *cardio-gram:* a record of the heart |
| **graph** | written record *encephalo-graph:* a written record of the inside of the brain |
| **gyn, gynaeco** | woman *gynaeco-mastia:* excessive development of male breasts |
| **haem, haemat, haemo** | blood *haemat-uria:* blood in the urine |
| **hemi** | half *hemi-plegia:* half of the body that is struck by a condition, stroke |
| **hom, homeo, homo** | same *homo-genous:* the same origin |
| **hydr, hydro** | water *hydro-phobia:* abnormal fear of water |
| **hyper** | above *hyper-thyroidism:* overactive thyroid gland |
| **hypo** | below *hypo-glossal:* below the tongue |
| **hyster** | womb *hyster-ectomy:* the removal of the womb |
| **iatro** | physician *iatro-genic:* originating from the physician |
| **in** | 1. in *in-fusion:* to run a liquid into the vein 2. negative *in-firm:* no longer strong |
| **infra** | below *infra-red:* below the wavelength of red light |
| **inter** | between *inter-costal:* between the ribs |
| **intra** | inside *intra-venous:* inside the vein |
| **is, iso** | equal *iso-metric:* equal length |
| **itis** | inflammation *nephr-itis:* inflammation of the kidney |
| **kin, kine** | movement *hyper-kinetic:* overactive movement |
| **lab, labi** | lip *labi-a majora:* large lips |
| **lact, lacto** | milk *lacto-genic:* causing the formation of milk |
| **larynge, laryngo** | windpipe *laryngo-scope:* to look at the larynx |
| **lent, lenti** | lentil-shaped *lenti-form:* formed like a lentil |
| **lep, leps, lept** | seizure *narco-leps-y:* the sudden onset of sleep |
| **leuco, leuko** | white *leuko-cyte:* white cell |
| **lingua** | tongue *lingua-l nerve:* nerve to the tongue |
| **lip** | fat *lip-oma:* a fatty swelling |
| **lymph** | water *lymph-oedema:* swelling due to water |
| **macro** | large *macro-cyte:* large cell |
| **mal** | bad *mal-ady:* illness |
| **man, manu** | hand *manu-al:* using the hand |
| **mani** | mental disorder *hypo-mani-c:* below the stage of mental illness |
| **mast** | breast *mast-ectomy:* removal of the breast |
| **mega, megalo** | great *megalo-mania:* mental disorder involving delusions of greatness |
| **men, mens, meno** | month *dys-meno-rrhoea:* disorder of the menstrual flow |

# Medical Terms

| element | meaning example definition |
|---|---|
| **metr, metro** | womb *endo-metr-itis:* inflammation of the lining of the womb |
| **micro** | small *micro-cephaly:* a small head |
| **mon, mono** | single *mon-ocular:* one eye |
| **mot** | movement *oculo-mot-or nerve:* a nerve that controls the eye muscle |
| **my, mya, myo** | muscle *my-algia:* pain in the muscle |
| **myel, myelo** | marrow, centre *myelo-cyte:* a cell from the marrow |
| **nas, naso** | nose *naso-pharynx:* nose and pharynx |
| **necro** | death *necro-tic:* concerned with death |
| **neo** | new *neo-plasm:* new shape |
| **nephr** | kidney *nephr-itis:* inflammation of the kidney |
| **neur, neuro** | nerve *neuro-logy:* the knowledge of the nervous system |
| **norm, normo** | usual *normo-chromic:* normal colour |
| **nutri** | nourish *mal-nutri-tion:* badly nourished |
| **ocul, oculo** | eye *ocul-ist:* eye specialist |
| **odont** | tooth *orth-odont-ist:* one who straightens teeth |
| **odyn** | pain *pleur-odyn-ia:* pain of the side |
| **oedem** | swelling *lymph-oedem-a:* swelling due to lymph |
| **oid** | form *andr-oid:* formed like a man |
| **olig, oligo** | small, few *oligo-menorrhoea:* a small amount of menstrual fluid |
| **oma** | swelling *oste-oma:* swelling of the bone |
| **oo** | of eggs *oo-phoritis:* inflammation of the ovary |
| **ophthalm, ophthalmo** | eye *ophthalmo-scope:* an instrument for looking at the eye |
| **or** | mouth *or-al:* concerning the mouth |
| **orchi** | testis *orchi-dectomy:* removal of the testis |
| **orth, ortho** | straight *orth-optics:* straightening of vision |
| **oss, osteo** | bone *oss-icle:* small bone |
| **ot, oti** | ear *ot-itis:* inflammation of the ear |
| **ov** | egg *ov-ary:* female sex organ where eggs are formed |
| **par** | bear *multi-par-ous:* one who has carried many children |
| **paed, ped** | child *paed-iatrics:* the treatment of disease in children *ortho-ped-ics:* the correction of deformities in children |
| **path, patho** | sickness *patho-logy:* the knowledge of sickness, in this case from studying dead tissue |
| **pen, penia** | lack *thrombocyto-penia:* lack of clotting cells |
| **peri** | around *peri-cardium:* that which is around the heart |
| **phag, phago** | eat *oeso-phag-us:* the gullet |
| **pharyng, pharynge** | throat *glosso-pharynge-al nerve:* the nerve to the tongue and throat |
| **phleb** | vein *phleb-itis:* inflammation of a vein |
| **phob** | fear *hydro-phob-ia:* fear of water |
| **phon, phono** | sound *phono-cardiograph:* writing from the sounds of the heart |
| **phor** | carrier *oo-phor-itis:* inflammation of the ovary |
| **plas, plasm, plast** | shape *cyto-plasm:* the shape of the contents of the cell |
| **plegi** | stroke *haemi-plegi-a:* a half-stroke |

| element | meaning example definition |
|---|---|
| **pleur, pleuro** | rib, side *pleur-odynia:* pain in the side |
| **pneum, pneumon** | lung *pneumon-ectomy:* removal of the lung |
| **poie** | produce *haemo-poie-tic:* blood producing |
| **poly** | many *poly-arthritis:* inflammation of many joints |
| **post** | behind *post-pharyngeal:* behind the pharynx |
| **pre** | before *pre-molar tooth:* tooth in front of the molar tooth |
| **pro** | before *pro-geria:* early ageing |
| **pseud, pseudo** | false *pseudo-cyesis:* false pregnancy |
| **psych, psycho** | mind *psycho-logy:* knowledge of the mind |
| **py, pyo** | pus *pyo-genic:* concerning the formation of pus |
| **radi, radio** | radiation *radio-logy:* study of radiation |
| **ren** | kidney *supra-ren-al gland:* the gland which rests above the kidney |
| **retro** | behind, backwards *retro-grade amnesia:* loss of memory of what occurred before an accident |
| **rhage** | flow *haemo-r-rhage:* flowing of blood |
| **rhin, rhino** | nose *rhin-itis:* inflammation of the nose |
| **rhoea** | gush *pyo-r-rhoea:* gushing of pus |
| **salping** | tube *salping-itis:* inflammation of the uterine tube |
| **sclero** | hard *arterio-sclero-sis:* hardening of the arteries |
| **scop** | look at *ophthalmo-scop-e:* an instrument to look at the eyes |
| **sect** | cut *dis-sect:* to cut apart |
| **semi** | half semi-circular canal |
| **sep, seps, sept** | decay *seps-is:* destructive infection of tissue by bacteria |
| **splanchn** | intestine *splanchn-ic artery:* the artery that goes to the intestine |
| **steat, steato** | fat *steato-rrhoea:* fatty diarrhoea |
| **stom** | mouth *colo-stom-y:* an opening, mouth, in the colon |
| **sub** | below *sub-mandibular:* below the mandible |
| **super, supra** | above *supra-umbilicus:* above the umbilicus |
| **sy, syl, sym, syn** | with *syn-chronous:* at the same time *sym-biosis* living together |
| **tachy** | fast *tachy-cardia:* fast heart-beat |
| **tele** | distant *tele-cardiography:* taking a cardiograph at a distance |
| **therap** | treatment *radio-therap-y:* treatment with radio-active substances or X -rays |
| **thorac, thoraco** | chest *thoraco-tomy:* cutting into the chest |
| **thrombo** | clot *thrombo-phlebitis:* inflammation and thrombosis of the veins |
| **tom** | cut *lobo-tom-y:* cutting of the frontal lobes |
| **tracheo** | windpipe *tracheo-stomy:* an opening, mouth, in the windpipe |
| **troph** | growth, nurture *dys-troph-y:* disorder of growth |
| **ultra** | beyond *ultra-sonic:* beyond the normal range of sound |
| **uri** | urine *poly-uri-a:* a condition of frequent passing of urine |
| **vesic** | bladder *retro-vesic-ular:* behind the bladder |

# Glossary

There are certain conventions when describing the position of a part of the body. It is assumed that the individual is standing upright with the palms of the hands facing forwards. This means that the diaphragm is superior to the liver and is inferior to the lungs, and the thumb is on the lateral side of the hand. There is a more detailed glossary of terms under the heading "Definitions" in each section of the book, from "How Cells Work" to "Metabolism and Nutrition". Diseases are provided with a brief definition under the heading "Common Diseases".

**Abdomen** Lower part of the trunk between the diaphragm and the pelvic girdle.

**Absorption** Process by which substances cross through the mucous membranes or skin. The products of digestion are absorbed from the small intestine into the blood-stream.

**Acoustic** Anything that is associated with sound or the sense of hearing.

**Adipose** Of a fatty nature.

**Adrenaline** Hormone produced by the medulla of the adrenal gland.

**Aetiology** Origin and cause of a disease.

**Agglutination** Clumping together of cells in fluid.

**Alimentary canal** Digestive tract, from the mouth to the anus.

**Allergy** Exaggerated response by the body's immune system to a specific external antibody, for example pollen, which does not produce a response in the majority of people.

**Amino-acid** One of the basic units from which proteins are made.

**Anabolism** Building of complex substances from simple components. The opposite to catabolism.

**Anaesthetics** Use of drugs to make a patient unaware.

**Anastomosis** Joining together of two pieces of tissue, for example, the joining of the jejunum to the stomach after a partial gastrectomy.

**Anatomy** Study and knowledge of the basic structure of man or animals.

**Androgen** Any substance that produces results similar to that of the male hormone.

**Anterior** Front or forward part of the body.

**Antibiotic** Substance either made from a fungus or living micro-organism or synthesized; it has the ability to kill or prevent the growth of pathogenic organisms.

**Antibody** Specific substance produced by the body as a reaction to and in an attempt to dispose of a foreign substance – antigen.

**Antigen** Any substance which, when introduced into the body, causes the formation of an antibody.

**Antiseptic** Any chemical used to destroy harmful micro-organisms.

**Antitoxin** Antibody that neutralizes bacterial poisons – toxins.

**Arteriosclerosis** Loss of elasticity of arterial walls with deposition of cholesterol in the degenerating areas.

**Arthrology** Study of joints.

**Articulation** The action or movement of a joint.

**Aseptic** Area that is free from infective organisms.

**Aspirate** To remove fluid or air from a cavity within the body.

**Auditory** Concerned with the mechanism of hearing.

**Bacteria** Microscopic organisms that are classified by their shape and staining capabilities. They live in many areas of the natural environment and only a few of them are pathogenic to the human body.

**Basal metabolism** Minimal rate at which the resting body uses energy.

**Benign** Not malignant.

**Bifurcation** Forking or division into two branches.

**Biochemistry** Study of the chemistry of living processes.

**Calibration** Measurement of an object.

**Calorie** Amount of heat required to raise 1 gramme of water through 1°C. The calorie value of food is the number of calories it would yield, as energy, if completely used. Nutritionally the Kilocalorie is used and is written as 1 Calorie, with a capital C.

**Cancer** Malignant growth in which the cells develop in a way that is no longer controlled and may spread throughout the body.

**Carbohydrate** Basic structure of starch or sugar.

**Carcinogen** Substance which will cause cancer to develop.

**Cardiology** Study of the heart.

**Catabolism** Breakdown of complex body substances into simpler compounds.

**Cellulose** Complex carbohydrate that forms a basic structural material of all plants.

**Cephalic** Concerned with the head.

**Cervical** Concerned with the neck of cervix.

**Cervix** Literally means neck, may mean cervix of uterus, tooth or bladder.

**Chromosome** Thread of genetic material. One of 23 pairs in the cell nucleus.

**Chyme** Liquid food in the stomach or small intestine.

**Cilia** Small hair-like structures on the outer surface of cells.

**Coitus** Sexual intercourse.

**Conception** Moment of fertilization.

**Consciousness** State of being awake.

**Corpuscle** See *Red blood corpuscle* and *White blood corpuscle*.

**Cortex** Outer layer, for example of the brain, adrenal gland or kidney.

**Cranial** Anything concerned with the skull.

**Cytology** Microscopic study of cells.

**Cytoplasm** Clear jelly-like substance that supports the intra-cellular contents.

**Decibel** Unit of measurement of sound.

**Dermatology** Study of skin disease.

**Diet** Food and drink that is consumed.

**Diffusion** Movement of molecules in a fluid from an area of high concentration to one of low concentration.

**Dilation** Stretching or widening.

**Dissemination** Scattering of a substance over a wide area, for example, cancer is disseminated throughout the body.

**Distal** Some distance from a certain place.

**Diuretic** Any substance that increases the output of urine.

**Dorsal** Anything that is related to the back.

# Glossary

**Embryo** Foetus in the early stages of development, before the fourth month.

**Endocrine gland** Gland that secretes a hormone directly into the blood stream.

**Endocrinology** Study of the endocrine glands.

**Enucleate** To remove an object from its place in the surrounding tissue without damage.

**Enzyme** Substance which speeds up chemical reactions.

**Exacerbation** Periods when a disease, or pain, becomes increasingly worse.

**Exocrine gland** Gland that secretes substances into a duct leading to the area where it is used; for example, the salivary gland is an exocrine gland.

**Faeces** Semi-liquid contents of the large intestine.

**Fascia** Any sheet of thin fibrous tissue.

**Fertilization** Moment when the sperm enters the ovum.

**Fissure** Term for any crack or cleft in the body.

**Foetus** Unborn infant from about the twelfth week of pregnancy until the time of birth.

**Frontal** That which is in front, particularly concerned with the forehead.

**Fructose** Simple sugar found in fruits.

**Gastric** Anything that is concerned with the stomach.

**Gastro-enterology** Study of the alimentary canal.

**Genital** Anything that is concerned with the sexual organs.

**Genito-urinary disease** Study of disease of the urinary and reproductive systems.

**Geriatrics** Care of the elderly.

**Gestation** Period of time between conception and birth.

**Gynaecology** Study of women's diseases.

**Haematology** Study of the blood cells.

**Haemopoiesis** Production of blood cells.

**Hepatic** Anything concerned with the liver.

**Heredity** Passage of family characteristics from parents to child.

**Homogenous** Same consistency all the way through.

**Hormone** Chemical produced in one part of the body, an endocrine gland, and carried to another area where it has a specific effect.

**Illusion** Incorrect reception of something actually present.

**Immunity** Body's ability to resist infection.

**Immunology** Study of the way the body reacts to disease.

**Infection** Invasion of the body by pathogenic organisms.

**Inferior** Situated below.

**Inflammation** Response of tissues to injury or infection. The signs are heat, redness, swelling and pain.

**Ingestion** Act of taking substances into an organism, for example, eating food or the ingestion of a harmful bacteria by a white blood corpuscle.

**Inhibition** Slowing down or stopping of a reaction, either chemical or neurological.

**Intelligence** Individual's capacity to think rationally, learn and react to the environment.

**Ischaemic** Insufficient blood supply to a part of the body.

**Isometric** Means the "same length" and is used medically to mean contraction of a muscle without bending a joint.

**Lateral** Concerning the side.

**Libido** Term used by Freud to describe the instinctive sexual drive in men and women.

**Malignant** Tendency to go from bad to worse, particularly used in reference to cancer.

**Mastication** Chewing of food.

**Maturation** State of full development.

**Medial** Towards the mid-line.

**Medulla** Term for the central part of an organ, for example, the medulla of the adrenal gland.

**Meiosis** Cell division in which there is a splitting to form a gamete with halving of each chromosome to form a chromatid.

**Menarche** Onset of menstruation at puberty.

**Menopause** Final cessation of menstruation in women usually occurring around the middle or late forties.

**Menstrual cycle** Sequel of events that occurs due to hormonal changes beginning and ending with the onset of menstruation. This usually takes about 28 days.

**Menstruation** Monthly "period" of uterine bleeding that occurs when the lining is shed.

**Metabolic rate** Rate at which an individual consumes energy reflected directly in the body's oxygen intake.

**Meticulous** With great care and precision.

**Micturition** Act of urination - passing water.

**Mitosis** Division of a cell to produce an identical one which contains the same number of chromosomes.

**Motile** Something that moves by itself.

**Mucous membrane** Thin, delicate lining, found in many of the body's internal surfaces, which is moistened with mucus secreted by the mucous glands in the membrane.

**Mucus** Semi-liquid substance secreted by the mucous glands in the mucous membranes.

**Myology** Study of muscles.

**Nasal** Concerning the nose.

**Necrosis** Death of a tissue.

**Neurology** Study of the nervous system.

**Neuropathy** Degeneration of a nerve.

**Obstetrics** Care and delivery of the pregnant woman.

**Occipital** Anything to do with the occiput – the back of the skull.

**Oestrogen** One of the group of female sex hormones which is principally produced by the ovary before ovulation but also from the adrenal glands.

**Olfactory** Anything concerning the sense of smell.

**Oncology** Study of cancer.

**Ophthalmology** Study of the eyes and sight.

**Optic** Anything to do with the eye and sight.

**Organism** Living thing, whether animal, plant or microbe.

**Orthopedics** Study of bone disease.

**Osmosis** Selective passage of the molecules of a liquid through a semipermeable membrane between solutions of different concentrations. Water from a weaker solution moves through the membrane to dilute a stronger one.

**Osmotic** pressure Pressure required to stop water from passing through a membrane into a solution; for example, osmotic pressure occurs when fluid is forced back into the capillary system after it has been forced out into the body tissues by the blood pressure.

**Osteology** Study of bones.

**Otorhinolaryngology** Study of the ear, nose and throat.

**Ovulation** Release of a mature ovum from an ovary.

**Paediatrics** Study of children's diseases.

**Palmar** Anything concerning the palm of the hand.

**Palpation** To examine a patient with the hand.

**Parturition** Act of giving birth to a child.

**Patent** Open.

**Pathogenic** That which gives origin to a disease, for example, a virus is the pathogenic organism of influenza.

**Peptone** Partly digested protein.

**Perception** Overall awareness of the environment from the interpretation of sensory information.

**Peripheral** Towards the outside.

**Pharmacology** Study of drugs in relation to medicine.

**Physiology** Understanding of the workings and functions of living organisms.

**Plantar** Concerning the sole of the foot.

**Posterior** Towards the back or situated behind.

**Pregnancy** Time from the moment of fertilization to parturition.

**Progesterone** Hormone produced by the ovary in the second part of the menstrual cycle to maintain the endometrium during implantation of the blastocyst.

**Pronation** Placing or holding of the palm of the hand downwards.

**Prophylactic** Something that is given to try and prevent a disease occurring.

**Protein** Complex nitrogen-containing substance concerned with building the body's fabric.

**Protuberance** Bulge or swelling.

**Proximal** Close to a certain place.

**Psychiatry** Study of mental illness.

**Puberty** Period during which the sexual organs reach maturity with the development of the secondary sexual characteristics, such as growth of hair, breast development and change of voice.

**Quickening** When the foetus is first felt to move by the mother is the onset of quickening.

**Radiology** Study of X-rays.

**Radiotherapy** Treatment of disease using X-rays.

**Red blood corpuscle** Small haemoglobin carrying cell that does not have a nucleus.

**Renal** Anything concerned with the kidney.

**Reproductive system** Sex organs concerned with producing sperm, the ovum and maintaining a pregnancy once it has started.

**Respiration** Process by which oxygen is delivered to the tissues and carbon dioxide removed.

**Retrosternal** Behind the sternum.

**Rheumatology** Study of the diseases of muscles and joints.

**Sac** Cavity within the body tissues, for example the pericardial sac.

**Sclerosing** Hardening of a tissue.

**Sensation** Experience of stimulation of a sensory nerve.

**Serum** Clear yellow fluid that separates from clotted blood.

**Skeleton** Articulated bony framework of the body.

**Smell** Response of the olfactory nerve endings to chemical stimulation.

**Stenosis** Narrowing, for example of an artery.

**Superior** Above or on top of.

**Supination** Placing or holding of the palm of the hand upwards.

**Synovial membrane** Smooth membrane that secretes a small amount of fluid and usually forms a sac to protect areas where friction may occur, for example joints and around tendons that pass over joints.

**Taste** Sensation of flavour that occurs when the sensory areas of the tongue are stimulated.

**Tendon** Strong, fibrous bundle which joins a muscle to a bone.

**Thorax** Chest compartment enclosed by the ribs, backbone and diaphragm. It contains the lungs, the heart, major blood vessels and the oesophagus.

**Threshold** Critical level of a stimulus to produce a response; for example, a feather lying on the skin may not be noticed, but a piece of gravel is above the threshold for feeling.

**Thrombosis** Formation or development of a clot of blood.

**Thyroid gland** Endocrine gland in the neck which produces hormones to stimulate the body's metabolism.

**Tissue** Large areas of similar cells, for example muscle tissue.

**Torsion** Act of twisting.

**Traction** Action of pulling on a muscle or organ by means of weights to correct an abnormal condition.

**Transient** Something that occurs only for a short time.

**Trauma** Wound or injury; emotional upset that may lead to mental illness.

**Tremor** Trembling.

**Umbilical** Concerning the navel.

**Unconscious** Absence of normal wakefulness, the lack of awareness that occurs when there is no response to normal stimuli. It is not, as a rule, applied to natural sleep.

**Venereology** Study of sexually transmitted diseases.

**Ventral** Concerning the abdomen.

**Virology** Study of virus disease.

**Virus** Complex chemical dependent on living cells for its reproduction.

**Viscus** Usually means the bowel, but can be referred to any of the abdominal organs, for example the liver and spleen.

**Vitamin** One of a series of substances that are found in the normal diet and are required in small amounts for the normal metabolism of the body and which the body cannot make from raw materials because it lacks the enzymes needed for their synthesis.

**White blood corpuscle** Concerned with fighting infection in the body.

# The Whole Body

**Looking at the body**

The systems that make the body work are all interdependent. This is sometimes difficult to remember when it is a matter of convenience to study each system separately. It makes learning about them easier, but it is just as possible to approach the body in a different, unconventional manner.

Little is known of how people develop as individuals with their own physical and psychological differences. Some of these are the result of inheritance; clearly much is due to immediate family, and from reactions to the wider world. Brain functions are notoriously difficult to measure; even intelligence tests are often innacurate and may provide little indication of the brain's power of original thinking.

However, the study of a particular protein will tell a great deal about how the body functions normally. Its reaction with the digestive juices breaks it into the constituent parts, amino-acids, which are then absorbed through the intestine wall and into the blood stream to the liver. In the liver the amino-acids may be rebuilt into new proteins or sent by the arterial system to help replace old proteins, such as those in muscle or bone. These "worn out" amino-acids are removed from the tissues by the venous system and returned to the liver. Here they can be rebuilt or broken down into urea; urea is excreted by the kidneys into the urine.

This unconventional way of learning teaches how the digestive tract, circulation, muscle and metabolism work, but could easily involve the hormones from the endocrine system, which help instruct the body, or the nervous system, where protein may be required.

A great deal is now known about how the brain controls the other systems of the body, either through quick direct nervous response or secretion of chemicals – hormones – to produce a slower reaction. The way in which the balance of salts in the blood is maintained by foods that are eaten is understood. All these physical occurrences can be watched and measured with accuracy.

# The Structure of the Body

### Definitions

■ **Achilles' tendon**
Large tendon from
the calf muscles to
the heel bone – the
calcaneus.

■ **Amino-acid** One of
the basic units from
which proteins are
made.

■ **Antibody** Specific
substance produced
by the body as a
reaction to, and an
attempt to dispose of,
a foreign substance –
antigen.

■ **Aponeurosis**
Tendinous expansion
connecting a muscle
with the parts that it
moves.

■ **Areolar tissue**
Loosely interwoven
connective tissue that
supports organs and
structures.

■ **Artery** Thick-
walled blood vessel
that carries blood,
under pressure, from
the heart.

■ **Cartilage** Special
tough form of
connective tissue; the
degree of hardness
depends on the
number of collagen
fibres.

■ **Collagen fibre**
Strongest form of
connective tissue.

■ **Connective tissue**
Supporting tissues of
the body.

■ **Elastic fibre** Elastic
form of connective
tissue.

■ **Fat cell** Cell
specialized for the
storage of fat to be
used as reserves for
either energy or
insulation.

■ **Ground substance**
Jelly-like material
made up of salt and
water, protein and
carbohydrate that fills
in the gaps of many
of the areas of
connective tissue.

■ **Inheritance**
Passage of family
characteristics that
are passed from
parents to child.

■ **Ligament** Tough,
fibrous band that
connects bones or
supporting organs.

■ **Macrophage**
Large tissue cell,
either free or
attached, which
engulfs and destroys
bacteria.

## A general description

The body as a whole is a compromise between rigidity and mobility. The internal organs (*right*) are closely packed together and yet can work freely and easily. The surrounding framework of bone and muscles applies support and protection.

The skeleton gives the upright strength to the body and in some places, such as the skull and thorax, acts as a protective layer. The joints give the bones mobility and the muscles strength and suppleness.

The contents of the chest and abdomen are constantly moving – the beating of the heart, inspiration and expiration in the lungs and peristalsis of the bowel. These structures can move without difficulty as they are surrounded by special, smooth layers of tissue known as pericardium, pleura and peritoneum. These small cavities of tissue, or sacs to give them their medical name, are rather like a very soft balloon containing a little fluid. Push a clenched fist in one side of a balloon, press the balloon against the palm of the other hand; the fist moves easily and without friction. Similar "sacs" occur in the joints, around tendons passing over joints and at the points of friction, such as the knee and elbow. These are known as synovial sacs.

The body's largest organ is the skin. In an adult it covers approximately 2 square metres and not only envelops the whole body in a protective waterproof layer but is also part of the heat-regulating system. The liver, meanwhile, is the most complicated organ with the greatest number of functions – transforming digested food into usable materials and disposing of waste substances.

The circulation is constantly restoring and revitalizing as well as removing waste products from the basic unit of the body – the cell. The cell is a microscopic structure of which there are billions that build up the whole body. Each cell specializes and carries out its own particular function. All the structures and organs are held together by the connective tissue, made up of cells that act as a kind of packing to protect and support the internal mechanisms.

## The internal organs

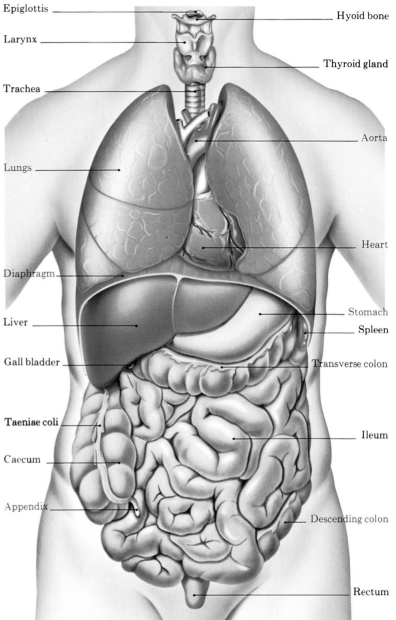

Epiglottis — Hyoid bone
Larynx —
Thyroid gland
Trachea —
Aorta
Lungs —
Heart
Diaphragm —
Stomach
Liver — Spleen
Gall bladder — Transverse colon
**Taeniae coli** — Ileum
Caecum —
Appendix — Descending colon
Rectum

**The contents of the body**
**The skeleton:** The axial and appendicular skeletons.
**Joints;** muscles; ligaments.
**Heart;** arteries, veins; blood; lymph vessels; bone marrow; spleen.
**Respiration:** Nose; sinuses; trachea; 2 bronchi; 2 lungs.
**Digestion:** Mouth; esophagus; stomach; small intestine; colon; liver; gall bladder; pancreas.
**Urinary system:** 2 kidneys; 2 ureters; bladder; urethra.

**Endocrine glands:** Pituitary; thyroid; 4 parathyroids; 2 adrenal; pancreas; 2 testes or ovaries.
**Nervous system:** 2 eyes, 2 ears; organs of hearing and balance, smell and taste; brain; spinal cord; peripheral nerves; autonomic nervous system.
**Skin, hair and nails.**
**Reproduction:** Female – 2 ovaries; 2 Fallopian tubes; uterus; vagina. Male – 2 testes; 2 epididymi; 2 sperm ducts; prostate.

## Connective tissue

Macrophage

Macrophage engulfing bacteria

Elastic fibres
Reticulin fibres

Plasma cell

Ground substance

Collagen fibre

Fibroblast
Arteriole
Fat cells

Nerve

## Connective tissue

The skeleton keeps the organs, blood vessels and nerves in place and, to a certain extent, affords some protection. The connective tissue supports and binds them together (*above*). It also supplies the ligaments and tendons for the joints and muscles, the tethering for the larger organs, the softness for protection and rigidity in the form of cartilage.

There are many forms of connective tissue, but they are all developed from the same jelly-like "ground substance", made up of salts and water, protein and carbohydrate. Embedded in this jelly are the various fibres and cells: elastic fibres to provide elasticity; collagen fibres to provide strength; reticulin fibres to provide support; white cells and macrophages to fight infection; fat cells used for storage; and finally plasma cells to produce antibodies.

## Collagen fibres – tendon

## Tendons and ligaments

The strongest connective tissue is found in the tendons and ligaments (*above*). These are made of densely packed collagen fibres and may be very thick, as in the Achilles' tendon, or thin and wide, as in the thin sheet of the aponeurosis that covers the skull and into which many of the skull muscles are inserted. They have to resist the pull of muscles and mobility of joints with suppleness and strength, without elasticity.

## Areolar

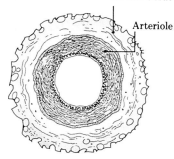

Areolar tissue

Arteriole

## Areolar tissue

The proportion of the different types of fibre in the "ground substance" gives the connective tissue its variable characteristics.

Areolar tissue is formed throughout the body in loose sheets around blood vessels, nerves and tendons; as a soft, pliable substance it helps to fill the spaces between larger organs. The tissue is made of a mixture of collagen, elastic and reticulin fibres. Under the skin and face it contains a large amount of mobile, elastic fibres, in contrast to the palms of the hands and soles of the feet, which are tough and contain more collagen fibres.

Connective tissue is found holding the cells together within organs. This can be seen in the liver, where the bile ducts, veins and arteries are held within the liver substance and in the artery wall itself; here the circular muscle is held in place by a protective sheath of loose areolar tissue (*above*). Certain diseases known as collagen diseases can afflict connective tissue, causing disorder of their normal function. The skin becomes wrinkled and creased with age as the elastic fibres degenerate. This occurs more rapidly when they are damaged by ultraviolet sunlight which is partly prevented by more pigmentation.

## Cartilage

Cartilage is a special form of connective tissue and supplies the fabric for the formation of bone. Bone forms by ossification – the process by which minute crystals of calcium salts are manufactured by osteoblast cells and arranged in layers.

Adult cartilage does not contain blood vessels or nerves but is filled with small holes to allow nutrition to seep into it. There are three forms of cartilage: Elastic cartilage is mainly densely packed cells to give it the kind of springiness found in the ear; fibro-cartilage is tough and contains many more collagen fibres (the intervertebral disc of the spine has a thick circle of fibro-cartilage around the softer centre of dense connective tissue, the nucleus pulposus); and hard hyaline cartilage, found at the bone ends and in the nose, is made of dense collagen fibre.

## Fat cells

## Fat cells

The fat cells (*above*) have three functions: Storage; insulation; protection over certain areas, such as the buttocks, and around various organs, such as the kidneys and heart and in the liver.

Some of the areas of the body consist mainly of fat storage cells. Fat cells develop in infancy and then their total number remains constant for the rest of life. Fat babies become fat adults.

# The Cell

## Definitions

■ **Adenosine triphosphate (ATP)** Chemical stored in the mitochondria that supplies energy for cell metabolism.

■ **Adrenal gland** One of the pair of endocrine glands lying above the kidney. It secretes corticosteroids from the cortex and adrenaline from the medulla.

■ **Adrenaline** Hormone secreted from the inner part of the medulla of the adrenal gland.

■ **Allele** One of a pair of chromosomes or genes.

■ **Carbohydrate** Basic structure of starch or sugar.

■ **"Carrier"** Individual who has a recessive gene for a condition, but does not develop it. Carried from one generation to another, the condition is only reproduced when it is paired with another recessive gene.

■ **Centriole** Areas within the cell that form part of the cell structure and also help in cell division.

■ **Chromatid** The "half" chromosome in the gamete when meiosis occurs.

■ **Chromatin** Central substance of a cell nucleus that forms chromosomes.

■ **Chromosome** Thread of genetic material. One of 23 pairs in the cell nucleus.

■ **Corticosteroid** One of several hormones secreted by the adrenal cortex.

■ **"Crossing over"** Method by which genes are interchanged between pairs of chromosomes before meiosis.

■ **Cytoplasm** Basic intracellular matter that supports the intracellular structures. It contains minute globules of fat and carbohydrates that combine with protein to form glycoprotein.

---

The cell is the basic unit of life. There are many varieties of single-celled organisms such as bacteria and amoeba. The human body is highly complex and contains approximately 10 million million cells which combine together to make the body function. All these millions of cells develop from an individual cell – the fertilized ovum.

Each cell is surrounded by a pliable and protective cell membrane (1). In the centre of the cell is the nucleus (16) which controls the cell's activities. The contents of the cell lie in the watery cytoplasm (4) which stores glycoprotein used for energy. In addition there are areas in the cell, the centrioles (6), which not only help to align the chromosomes when the cell divides but also form filaments of protein, the cytoplasmic network, (5) that provide rigidity, like a skeleton, for the cytoplasm.

### The cytoplasm

Substances entering the cell and also degenerating parts of the cell itself are broken down by the enzymes in special areas called lysosomes (12). Films of material within the cytoplasm, the endoplasmic reticulum (11), form sacs and tubes connected to the nuclear envelope (15). Ribosomes (7) covering this reticulum are the cell's production areas and are under the control of the ribonucleic acid (RNA) from the nucleoli (13).

The energy for all this activity is stored in the many minute mitochondria (3) in the form of adenosine triphosphate (ATP). The substances, enzymes, hormones and other chemicals, are transported to the Golgi complex (10). The Golgi complex modifies and sorts them into globules (9) that are secreted through the cell membrane (8) to be removed by the circulation.

### The cell nucleus

The nucleus (16) controls and directs the activities of the cell. It is surrounded by its own nuclear envelope (15) which has pores (14) to allow chemicals to flow in and out. It is filled with densely packed chromatin which unfolds, when the cell divides, to form chromosomes containing deoxyribonucleic acid (DNA) and RNA. The 46 human chromosomes contain the genes that determine the activity of each cell and the body as a whole. In the nucleus are the nucleoli (13), usually between 1 and 4 in number depending on the type of cell. The nuclei are composed of RNA which helps to control the activity of the ribosomes (7).

### The cell as a factory

The metabolism of the cell depends on a constant supply of raw materials and the removal of the finished substances and waste by the blood circulation. In this way each cell is like a factory. There are three stages: raw materials; manufacture; and disposal. It is controlled by the nucleus and maintained by the energy reserves.

### The cell membrane

Selected substances can pass through the cell membrane in both directions (2 and 9). The membrane (Fig.2) is a double layer of lipid molecules (a) in which there are special tubules (b) and chemicals that aid salts, sugars and aminoacids in their transport into the cytoplasm. Other surface proteins (c) are concerned with immunity and the way that cells recognize each other. Cholesterol molecules (d) help to keep the membrane in good condition.

Substances leaving the cell do so by forming globules (3) and coalescing with the membrane after leaving the Golgi complex (10), a process known as exocytosis, and are then extruded into the interstitial fluid.

---

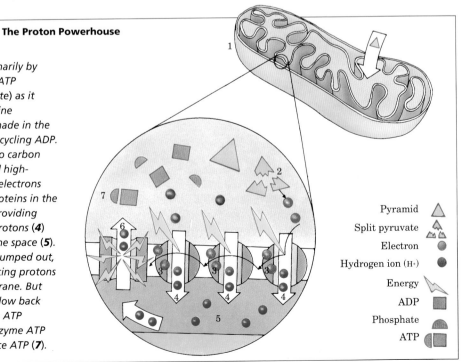

**The Proton Powerhouse**

*Cells are powered primarily by energy released from ATP (adenosine triphosphate) as it becomes ADP (adenosine diphosphate). ATP is made in the mitochondria (1) by recycling ADP. Pyruvate (2) is split into carbon dioxide, hydrogen and high-energy electrons. The electrons pass along a line of proteins in the inner membrane (3) providing energy to pump out protons (4) into the intermembrane space (5). As more protons are pumped out, pressure builds up forcing protons back across the membrane. But the protons can only flow back into the matrix via the ATP generator (6) – the enzyme ATP synthetase and produce ATP (7).*

Pyramid

Split pyruvate

Electron

Hydrogen ion (H+)

Energy

ADP

Phosphate

ATP

**Basic animal cell**

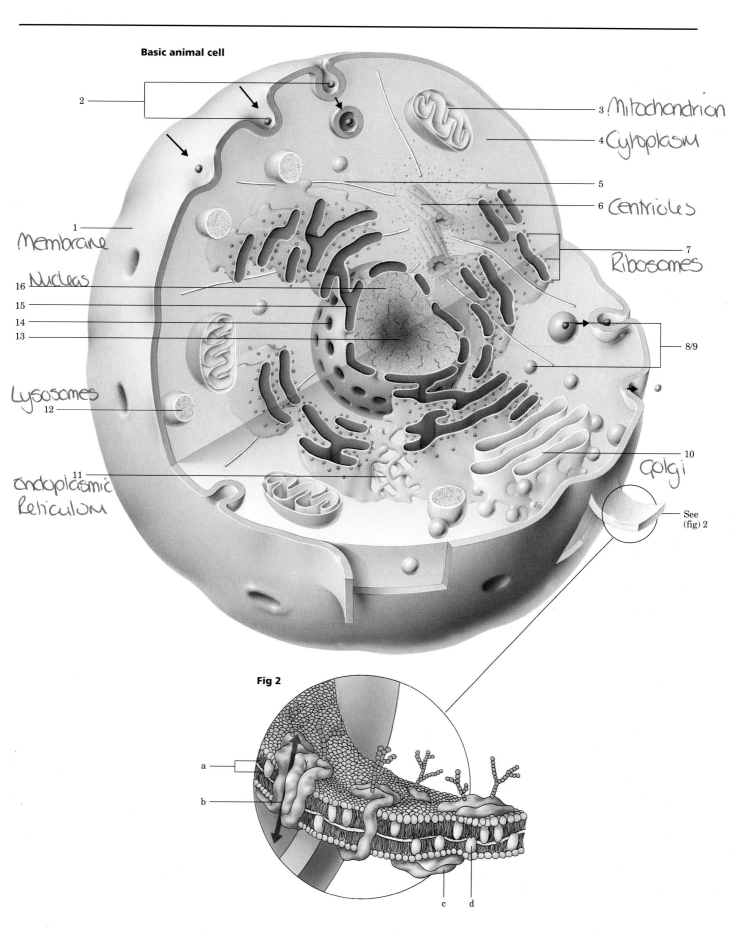

2

3 Mitochondrion

4 Cytoplasm

5

6 Centrioles

1 Membrane

7 Ribosomes

16 Nucleus

15

14

13

8/9

Lysosomes

12

10

11 Golgi

endoplasmic Reticulum

See (fig 2)

**Fig 2**

a

b

c   d

15

# How Cells Work

■ **Cytoplasmic network** Faint strands of material that form in the cytoplasm and which pull the chromosomes into the sides of the cell during cellular division.

■ **Deoxyribonucleic acid (DNA)** Complex chemical that is the foundation for the memory of the gene. It produces ribonucleic acid (RNA).

■ **Dominant** The stronger.

■ **Endoplasmic reticulum** Film of intracellular matter to which the ribosomes are attached and on which most of the enzyme activities of the cell take place under the direction of ribonucleic acid.

■ **Enzyme** Chemical that helps the transformation of one substance into another – a specific catalyst.

■ **Epithelium** Layer of cells on the outside of a structure, e.g. skin, or lining the bronchi.

■ **Gamete** Cell containing half the number of chromosomes (23) after meiotic cellular division.

■ **Gene** Basic unit of a chromosome which carries instructions in the DNA for one body characteristic.

■ **Genotype** Combination of genes that produces the same characteristics in an individual.

■ **Glycoprotein** Substance formed from the combination of a protein with fat and carbohydrates. It is stored in the cytoplasm.

■ **Golgi complex** Intracellular structures of the cell that control the enzyme activities of the endoplasmic reticulum.

■ **Helix** Anything with a spiral form, hence the double helix of the spiralling appearance of DNA.

■ **Heterozygous** Chromosomes containing different genes.

## The three stages of production in a cell

Raw materials: Each cell type requires different substances and these can only be obtained from the circulation. They are carbohydrates, fats, amino-acids and various salts, which enter the cell through the cell membrane. Depending on the cell's function, only certain substances are allowed through.

Manufacturing: The process takes place on the surface of the endoplasmic reticulum found throughout the cytoplasm. Some of the products may be enzymes or hormones.

Disposal: The final product and the waste substances, which are a result of the cell's activities, are passed through the cell wall into the interstitial fluid and then into the blood for disposal.

## The cell's energy reserves

Potential energy is stored in the mitochondria as glucose, fat globules and protein and is released in the form of the chemical adenosine triphosphate (ATP) when it reacts with oxygen. This releases carbon dioxide and water as waste products.

**Chemical equilibrium in the cell**

### Epithelial cells

Photo: Ron Boardman

### The cell's control

Control for all this activity comes from the nucleus, a memory bank for everything concerned with the cell's functions. The genes in the chromosomes are each a deoxyribonucleic acid (DNA) molecule. They send out chemical messengers in various forms of ribonucleic acid (RNA) to construct the amino-acids into the precise chains to form proteins in the cytoplasm – the material that forms the bulk of the cell.

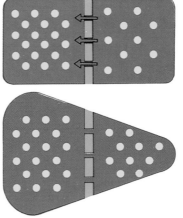

### Chemical equilibrium in the cell

Diffusion (*top left*) is the passive movement of molecules from high to low concentration. Active transport (*left*) requires energy. The molecules are "bound" by chemicals on one side of the wall and taken through to the other side. Osmosis (*above*) is the movement of water through the semi-permeable cell wall, from a weaker to a stronger solution.

### Simple and complex cells

The epithelial cells of the tongue are shown (*above*) magnified 500 times. They are among the simplest cells in the body, with a tough, resistant cell membrane surrounding the cytoplasm and nucleus. They are layered together and firmly attached to the underlying tissue. Nutrition reaches them by diffusion through the interstitial spaces and their only function is to replace themselves as a firm, waterproof layer.

Other simple cells are found in the skin or the nails and hair, where substances like keratin are produced.

Within the body more complicated cellular processes occur involving chemical excretions; for example digestive enzymes from the pancreas and hydrochloric acid in the stomach. Every minute 300 million cells in the body die and are replaced immediately by division of the living cells, so that the number of cells in the adult body remains constant throughout life.

### Hormone control

The circulation brings hormones such a thyroxine, and the corticosteroids from the adrenal gland; these help regulate the speed of the cell's activity. Some hormones have a specific effect on one type of cell, for example thyroid-stimulating hormone from the anterior pituitary gland, which affects only the thyroid gland. Other hormones affect one type of activity in all cells, such as insulin from the pancreas, affecting carbohydrate metabolism.

## Chromosomes

Every human cell contains 46 chromosomes (*right*), except the sperm and ova, which have only 23. The chromosomes are compressed into the nucleus as chromatin. The sex chromosomes called X (red) and Y (blue) are circled. A combination of the longer X and shorter Y denotes male characteristics, an XX pair those of the female.

Each chromosome consists of a double strand of deoxyribonucleic acid (DNA) molecules. Each huge molecule of DNA is a gene. The combination of about 500 genes on each of the 46 chromosomes carries sufficient instructions for all the activities within the body, the way it is built and all the inherited instincts, from the colour and size of our eyes to the speed of our reactions.

## Forty-six chromosomes

Photo: Ken Moreman

## Cell division – mitosis

During mitosis (*right*), chromosomes become shorter (**1**) and the nuclear envelope breaks (**2**), releasing the chromosomes, which duplicate and attach themselves to a cytoplasmic network (**3**). They are then drawn apart (**4 – 6**) to form two new cells with re-formed nuclear envelopes (**7**).

## Mitosis – cell reproduction

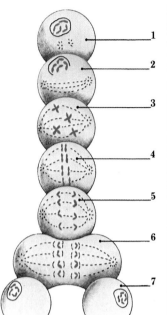

## DNA – the code of life

**Model of DNA**

Photo: Courtesy Prof. M. H. F. Wilkins, King's College

## DNA – the code of life

DNA is the basic unit of control of human life. It is a highly complex substance formed from a chain (*above*) of chemical units known as nucleotides. Each nucleotide is made from a sugar, phosphate and one of four different kinds of nitrogen-containing compounds called bases. The sugar and phosphate form the side of a ladder and the bases link in the middle to form the rungs of this spiralling ladder (**1**), the so-called double helix. These are the genes; it is the varying chemical composition of the DNA of each that causes the different RNA production.

When a cell is about to divide, the DNA molecules reproduce themselves by doubling and splitting lengthwise. When the nuclear envelope breaks, the two identical halves attach themselves to a cytoplasmic network that divides them.

DNA works by a temporary break in the double helix and the production of RNA, a sugar-phosphate-base complex similar but slightly different to the DNA. The RNA leaves the nucleus and acts as a guide for protein formation by the ribosomes on the endoplasmic reticulum. This new protein may be used elsewhere in the body or during the duplication of the DNA in mitosis.

The gradual magnification of a simple gene (**2**) to the final molecular structure of DNA illustrates its complexity. The model (*right*) is only a tiny portion of the whole molecule. The different bases are shown in varying colours.

# The Cell – Inheritance and Genes

■ **Ovum** Female gamete.

■ **Pancreas** Digestive organ that produces enzymes and the hormones insulin and glucagon.

■ **Pituitary gland** Most important endocrine gland, situated under the brain. It has two parts, the anterior lobe and posterior lobe.

■ **Recessive** Weak.

■ **Ribonucleic acid (RNA)** Chemical produced by DNA that controls the cell's activity by reproducing proteins on the ribosome.

■ **Ribosome** Area on the endoplasmic reticulum where the RNA reproduces proteins.

■ **Sex-linked** Genes that are found on the sex chromosomes.

■ **Sperm** Male gamete.

■ **Thyroid gland** Endocrine gland in the neck that produces thyroxine.

■ **Thyroid-stimulating hormone** Hormone produced by the anterior pituitary lobe that controls activity of the thyroid gland.

■ **Thyroxine** Hormone produced by the thyroid gland that helps control metabolism of the body.

■ **Zygote** New cell, containing 46 chromosomes, which is formed by the joining together of two gametes, each containing 23 chromatids.

## Heredity

Normally a cell divides and the two new cells contain the same number of identical chromosomes. This is mitotic division. Human cells contain 23 pairs of chromosomes.

Reproductive cells – gametes – are formed by the splitting of the chromosome pairs to produce 23 chromatids in each cell. This is meiotic division. If this were the only thing to happen all the children of any one couple would be identical.

Before meiosis occurs an apparently normal mitotic cell division takes place, but the chromosomes overlap and seem to "stick" to each other. When they are pulled apart by the cytoplasmic network, the chromosomes break and change sections with each other, thus altering the genes. This is known as "crossing over" (*lower right*).

As each gamete is different, this alteration of genes gives each person a unique individuality. There are many features that will be nearly identical in close relatives and this makes members of the family recognizable as relations. Identical twins occur when the zygote – fertilized ovum – splits mitotically, producing two identical cells.

When the chromatids meet to form new pairs – alleles – certain physical features are opposite each other and, like the chromosome, become paired. There are probably 50,000 pairs of genes in each zygote carrying all the future instructions for the body. Many of these pairs will carry identical instructions – known as homozygous – but if they are different – heterozygous – one gene will have "stronger" instructions than the other. This is known as the dominant gene. The weaker one is called the recessive gene. Some genes are not always dominant or recessive; one may only tend to dominate.

The majority of genes do not vary and are homozygous. Most people have the same number of features and grow in a very similar manner. If the genes are badly damaged it is unlikely that the zygote will develop at all. Minor genetic changes may produce abnormalities, usually harmful but occasionally beneficial.

## Meiosis – gamete formation

## Chromosomes crossing over

## Gamete formation

Gamete cells are produced to transmit genetic information (*above*) from parents to their children.

The chromosomes appear in the nucleus (**1**).

The DNA splits and the chromosomes are duplicated as the nuclear envelope breaks (**2**).

Similar chromosomes overlap and adhere to each other – "crossing over" – exchanging sections (**3**).

The cytoplasmic network pulls the chromosomes apart to line them up in the centre (**4**).

The chromosomes are then pulled to opposite poles of the cell (**5**).

The cell membrane breaks and two new cells are formed, each containing 46 non-identical chromosomes (**6**).

Meiosis can now start (**7**).

The two chromosome pairs split apart and each half, a chromatid, is pulled by the cytoplasmic network to the side of the cell (**8**).

The cell membrane breaks and the two new pairs of cells are formed, each containing 23 chromatids. This is a gamete. The joining of two gametes causes pairing of the chromatids and formation of a new cell, the zygote, with 46 chromosomes (**9**). Occasionally more or less than 46 chromosomes are formed, which usually causes a severe disorder.

## Haemophilia – European royal families

female carrier | male afflicted | female non carrier | possible female carrier | possible male afflicted | male unafflicted

Edward, Duke of Kent (1767–1820) — Victoria, Princess of Saxe–Coburg (1786–1861)

Albert, Prince of Saxe–Coburg–Gotha (1819–61) — Queen Victoria (1819–1901)

Leopold, Duke of Albany (1853–84) — Helen (1861–1922); Louis IV. (1837–92) — Alice (1843–78); Edward VII, King of England (1841–1910); Victoria (1840–1901) — Frederick III, Emperor of Germany (1831–88); Beatrice (1857–1944) — Henry, Prince of Battenburg (1858–96)

Alice (1883– ); Alix, Czarina of Russia (1872–1918); Frederick (1870–73); Irene (1866–1953); George V, King of England (1865 1936); Sigismund (1864–66); Waldemar (1868 79); Victoria, of Spain (1887–1969); Leopold (1889 1922); Maurice (1891 1914)

Mary (1906– ); Rupert (1907 28); Maurice (b.d. 1910); Olga (1895–1918); Tatiana (1897–1918); Marie (1899–1918); Anastasia (1901–1918); Alexis (1904–18); Waldemar 1899–1945; Henry (1900–04); Alfonso (1907–38); Beatrix (b.d. 1910); Maria (1911– ); Gonzalo (1914–34)

Anne (1932– ); Elizabeth (1936– ); Present British Royal Family; Sandra (1936– ); Olympia (1943 ); Victoria (1941– ); Giovanna (1943– ); Maria (1945– ); Anna (1948– )

## Hereditary defects

Many of an individual's characteristics are due to the balance in his genetic inheritance. Most characteristics are formed by more than one gene, such as eye colour; others are produced by the dominance of one gene only, such as that in blood groups and sub-groups. Rare conditions are due to the combination of two recessive genes; this can be seen in albinism – lack of the pigment melanin – when both parents may be "carriers" without any sign of the condition.

Sex-linked conditions are due to the gene being absent in the smaller male allele of the pair of sex chromosomes, so that the gene on the female half asserts itself. Haemophilia, an inherited deficiency in a blood-clotting factor, is such a condition. From the family tree of the royal families of Europe (*above*), it seems probable that Queen Victoria spontaneously produced the gene for haemophilia. Colour-blindness is another sex-linked condition and is ten times more common in males than females. As a colour-blind man may marry a "carrier" some of the daughters in turn may also be colour-blind.

## Sex of an individual

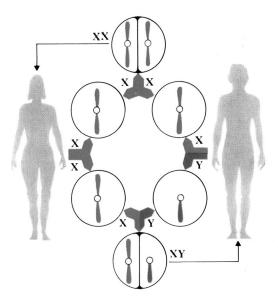

### The sex of an individual

An individual's sex is determined by a particular pair of chromosomes (*above*). If they are homozygous a female is produced and the two alleles are of equal length with the same number of genes. These are called XX. If the pair is heterozygous, XY, a male is produced. The Y allele is smaller and contains fewer genes than the X half of the pair. Thus the unpaired genes act in a dominant manner.

## Eye colour

A    B

### Eye colour

There is more than one gene for eye colour (*above*), but brown is dominant over blue. Two people, one with two genes for brown eyes, the other with two genes for blue eyes (**A**), will have children who all have brown eyes. However, if two brown-eyed parents (**B**) carry the heterozygous recessive blue gene they will have one blue-eyed child for every three brown-eyed children.

**Cross-references**

Vision 67
Childbirth 84-5

# The Skeleton

### Basic facts

The skeleton consists of about 206 bones divided into two broad groups, the axial and appendicular skeletons. It has three functions: It supplies support; it protects the internal organs; and, by using muscles, it gives movement. The axial skeleton, consisting of the skull, spine and rib cage, supplies the basic structure on to which the limbs, the appendicular skeleton, are joined via the pelvic and shoulder girdles.

### Appendicular skeleton – upper limb

The shoulder girdle consists of two bones, the clavicle and scapula. The clavicle acts as a strut with one end fixed against the manubrium sterni while the outer end holds the scapula from the thorax. The glenoid cavity is for the joint with the humerus.

The arm consists of long bones and a highly mobile hand. The humerus joins the radius and ulna at the elbow and the mobility of these two bones allows pronation and supination of the forearm and hand.

### Appendicular skeleton – lower limb

The pelvis is much stronger than the shoulder girdle because it has to support the full weight of the body. Each innominate is formed from 3 bones: The wing-shaped ilium; the pubis in front; and the ischium behind. These are fused together. In front the innominate bones articulate at the symphysis pubis and join the massive sacrum behind. All these 3 bones form the acetabulum for the articulation of the femur.

The lower and upper limbs are similar but with different functions. The long femur has its head extended to the side by the neck to articulate in the acetabulum. Large tubercles – the greater and lesser trochanters – are formed for muscle insertion. The lower end of the femur articulates the tibia, with the patella in front; the fibula supplies support for the muscles and part of the ankle joint. The 7 tarsals and 5 metatarsals support the body weight; the 14 phalanges are much smaller in the hand as they have little active function.

Phalanges

Parietal
Frontal
Nasal
Temporal
Zygomatic
Maxilla
Mandible
Cervical vertebrae
Clavicle
Sternum
Ribs
Xiphisternum
Humerus
Radius
Ulna
Ilium
Pubis
Femur
Patella
Fibula
Tibia
Tarsals
Metatarsals
Phalanges

Parietal

Occipital

Cervical vertebrae

Clavicle

Scapula

Ribs

Humerus

Radius
Ulna

Femur

Fibula
Tibia

Metacarpals
Carpals

Atlas
Axis

Thoracic vertebrae

Lumbar vertebrae

Sacrum

Coccyx
Ischium

Calcaneus

## Axial skeleton – spine
The spine consists of 7 cervical, 12 thoracic and 5 lumbar vertebrae. The 5 sacral and 4 vertebrae of the coccyx fuse together to make solid bone.

## Axial skeleton – thorax
The thorax consists of 12 pairs of ribs, articulating with the thoracic vertebrae, 10 pairs joined with cartilaginous processes to the sternum in front leaving the two lowest pairs "floating". The sternum consists of the manubrium, at the top, and the small xiphi-sternum at the lower end.

Some people have an extra vertebra or an extra rib.

## Bone structure
All bones have an outer, compact, dense layer and inner, spongy, cancellous centre. This makes them strong but light. Bones also store calcium and phosphorus and in many the cancellous centre is replaced by a medulla containing the marrow – blood-forming cells – or, in the case of sinuses, an air-containing space.

The articular surfaces of bone are covered with cartilage to supply a smooth surface for the joint. The overall surface of most bones is irregular and grooved by blood vessels or nerves. Bones are surrounded by tough, fibrous periosteum into which muscles and ligaments are inserted. Their traction causes ridges, tubercles or crests to occur by periosteal reaction – new bone formation. There is no nerve supply to bone, but blood vessels enter through the nutrient canal to reach the cancellous centre.

## Bone growth
Growth takes place in all bones, but is more obviously apparent in the long bones. An infant's long bones have bony ends – epiphyses – separated from the shaft by the cartilagenous diaphyseal line of new bone formation. By the age of 25 (earlier in women) all these diaphyseal lines have ceased growing and have fused with the bone. Short and irregular bones, like carpal bones and the maxilla, calcify from the centre. Flat bones start as a membranous sheet that forms bone and gradually thickens to a centre of cancellous bone.

### Definitions

■ **Acetabulum** Cup-shaped cavity on the side of the pelvis in which the head of the femur articulates.

■ **Calcify** Deposition of calcium salts in a tissue.

■ **Cancellous** "Honeycomb" bone structure.

■ **Cartilage** Special, tough, firm form of connective tissue, the degree of hardening depending on the number of collagen fibres.

■ **Cervical** Concerning the neck.

■ **Collagen** Tough connective tissue.

■ **Conchae** Three shell-shaped structures on the lateral wall of each nasal passage, 2 of cartilage and 1 of bone.

■ **Cranium** Bones of the skull that protect the brain.

■ **Diaphysis** Cartilaginous strip separating the epiphysis from the shaft to the long bone which is concerned with growth.

■ **Epiphysis** Piece of bone separated from the shaft of the long bone by the diaphysis.

■ **Fontanelle** Unossified area, between bones, in the skull of an infant.

■ **Foramen** Hole in a bone.

■ **Glenoid cavity** Area of articulation between the humerus and scapula.

■ **Haversian system** Tubular system of which bone is formed. A central canal surrounded by calcified tissue.

■ **Interstitial fluid** Body fluid outside the circulation that bathes the cells.

■ **Lumbar** Concerned with the region of the lower back.

■ **Meninges** Membranes that surround the brain and spinal cord.

■ **Osteoblast** Bone cell that deposits calcium salts in the cartilage.

# Specialized Bone

## Bone formation

All bones are formed from cartilage, with the exception of the clavicle and some of the skull, which ossify directly from membrane.

Compact bone is surrounded by periosteum and made of long columns of calcified tissue. These columns are known as Haversian systems (*right*), each with a central canal containing an artery, vein and lymph vessels, and filled with interstitial fluid. Osteoblasts – bone-forming cells – lay down concentric rings – lamellae – of mineral salts, mainly calcium and phosphorus, in the framework of the cartilage. These layers have collagen fibres running in different directions to each other to give greater strength. Long bones are also formed from the periosteum.

In cancellous bone the Haversian systems are less densely packed and the collagen fibres are in a honeycomb network pattern. This gives lightness in contrast to the strength from compact bone.

**Bone formation**

Blood and lymph vessels in central canal

Haversian system

Lamellae of calcium salts

Periosteum

Section through **A**

## The hand

Distal phalanx

Middle phalanx

Proximal phalanx

5th Metacarpal

1st Metacarpal

Trapezoid
Trapezium
Capitate
Scapmoid

Hamate
Pisiform
Triquetrac
Lunate

## The foot and the hand

Although they have very different uses the foot and the hand are similar in structure. The foot bears the full weight of the body and acts as a spring when walking and running. The massive calcaneum thrusts the weight on to the ground at the back, this is transferred to the front of the foot by the higher medial and smaller lateral longitudinal arches to reach the bases of the first and fifth metatarsal bones. There is a smaller transverse arch between these two points that turns the foot into a triangle of arches held together by the ligaments, tendons and muscles of the foot.

The hand (*left*) is far more delicate, with long, easily mobile fingers, each containing 3 phalanges. A thumb, of 2 phalanges, has the unique ability to reach across the palm, made from 5 metacarpals, thus making grasping possible. The mobility of the 8 carpal bones in the wrist, with the fingers and thumb, has produced an instrument, associated with the eyes and brain, able to carry out skilled and accurate movements in a way that would be impossible with the foot.

**Exploded view of the skull**

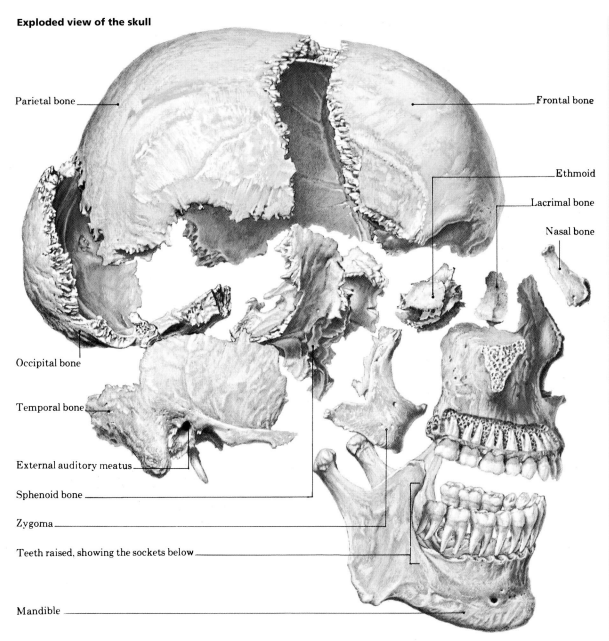

Parietal bone

Frontal bone

Ethmoid

Lacrimal bone

Nasal bone

Occipital bone

Temporal bone

External auditory meatus

Sphenoid bone

Zygoma

Teeth raised, showing the sockets below

Mandible

## The skull

The skull (*above*) is both the shell of protective bone for the brain, eyes and organs of hearing, and the structure of the face.

The eight bones of the cranium – occipital; 2 temporal; 2 parietal; frontal; sphenoid; ethmoid – are fused together. This rigid structure is pierced by foramina for the cranial nerves and blood vessels. It is marked internally by the meningeal arteries and externally by ridges at muscle insertions. The occipital bone contains the foramen magnum and articulates with the atlas. The sphenoid bone contains a cup-like cavity to protect the pituitary gland. The 3 bones of the middle ear are found deep inside each temporal bone. In infants, before the bones fuse, fontanelles are found anteriorly between the parietal and frontal bones and posteriorly between the parietal and occipital bones.

The face is composed of 14 bones: 2 maxillae; 2 nasal; 2 zygoma; 2 lacrimal; the mandible and (not shown in the picture) 2 palatine; 2 inferior turbinate bones (the superior and middle nasal conchae are cartilagenous) and vomer (the inferior and posterior parts of the nasal septum).

The mandible, like the maxilla, contains sockets for the 32 teeth, which are embedded in fibrous tissue. The nerves from the organ of smell – cranial nerves – penetrate the nasal part of the ethmoid bone. The frontal, maxillary, ethmoid and mastoid process of the temporal bones contain air sinuses to lighten the skull.

**Cross-references**

Whole body 10–11
Joints 24–5
Blood 34–5

# Joints

## Basic facts

A joint is the meeting point between bones; it usually allows a controlled amount of movement. Some joints have to be very strong, while others have to be very mobile. It is not possible for a joint to be both strong and mobile. Joints can be classified according to their mobility:

## Mobile joints

Because they have to be able to withstand the friction of movement these joints (below and right) are complex. The surfaces of the bone are covered with smooth cartilage. The joint edge has a strong fibrous capsule surrounding a sac of synovial membrane between the bone ends. This membrane secretes lubricant fluid that allows frictionless movement and the articular cartilage keeps the bones apart. The joint has stabilizing ligaments which bind and strengthen it.

## Slightly mobile joints

Some joints require a small amount of movement, but still have to be very strong. This can be seen between the vertebrae of the spine and the symphysis pubis. In these joints there is a thick pad of fibro-cartilage between the bones held in place by a strong fibrous ligament. This pad acts as a shock absorber.

## Immobile joints

In infancy many of the skull bones are not joined together and in two places, the anterior and posterior fontanelles, there are gaps between them. As the infant grows these open joints join to form the rigid skull. The irregular, serrated edges of bone are bound together by tough, fibrous tissue at the suture lines.

**Mobile joints**

**Ellipsoid joint** Allows circular and bending movement but no rotation, e.g. between finger and palm.

**Saddle joint** Allows movement in two directions, but without rotation, e.g. ankle, and thumb joints between wrist bones and thumb.

**Ball-and-socket joint** A joint freely moving in all directions, e.g. shoulder and hip joints.

**Hinge joint** Allows extension and flexion, e.g. elbow and finger joints.

**Condylar joint** This is similar to a hinge joint, but with slight rotation to allow the joint to "lock" into an extended position, e.g. the knee joint.

**Plane joint** A flat surface allows the bones to slide on each other, but they are restricted by ligaments to a small range, e.g. tarsal bones of foot and between ribs and thoracic vertebrae.

**Pivot joint** Allows rotation, but no other movement, e.g. between the atlas and axis.

## The spine

## The spine – discs

The main joints between the vertebrae of the spine (*left*) are slightly mobile. The vertebral surface (*below*) is covered with hyaline cartilage (**1**) and the intervening space is filled with a thick ring of fibro-cartilage (**2**) with a centre of soft, almost gelatinous, tissue – the nucleus pulposus. The joints are held together by the anterior (**3**) and posterior (**4**) longitudinal ligaments, the ligamentum flavum (**5**) in the neural canal (**6**), the interspinous (**7**) and supraspinous (**8**) ligaments.

Neural foramina
Anterior longitudinal ligament
Posterior longitudinal ligament
Hyaline cartilage
Intervertebral disc, fibro-cartilage
Neural canal
Interspinous ligament
Ligamentum flavum
Supraspinous ligament

## The spine – movement

The spinal vertebrae also have joints between their other articulating surfaces, on the neural arches, and with the ribs in the thoracic region. These have a synovial membrane and are surrounded by ligaments. This allows a much greater degree of movement. The joints between the atlas and occiput and between the axis and atlas do not have discs but rely on synovial membranes to give freedom of movement.

The intervertebral discs act as "shock absorbers" for the skull and brain. The movement between individual vertebrae, with the exception of the axis and atlas, is small, but the overall combined effect is considerable. Most of the flexion and extension is in the cervical and lumbar regions, while bending to the side is principally in the thoracic area. Twisting involves the whole vertebral column. Local movement allows the nerves to slide out of the neural canal through the neural foramina (**9**), without damage.

## Ball-and-socket joints

In the hip joint the head of the femur fits into the acetabulum and is held in place by a rim of cartilage. Inside the joint the strong ligamentum teres holds the femur to the pelvis.

The shoulder joint is similar, being held in place by the surrounding muscle tone as well as the ligaments. However, it is relatively weak and has a much shallower joint.

## The knee joint

The knee joint (*right*) is the largest and most complex joint in the body. As it reaches full extension it rotates slightly and "locks" into a rigid limb from hip to ankle, helped by the bony protuberances at the lower end of the femur and the two semi-lunar cartilages attached to the sides of the upper end of the tibia. The ligaments surrounding the joint are combined with the cruciate ligaments to prevent over-extension. The tendon of the quadriceps femoris muscle surrounds the patella bone in front of the knee. This helps to protect the joint. Behind some tendons and ligaments there are bursae.

## The knee joint

Cruciate ligament
Medial ligament
Medial cartilage (meniscus)
Patellar ligament
Tibia
Patella
Lateral cartilage (meniscus)
Lateral ligament
Fibula

■ **Arthritis** Inflammation of a joint. More often due to autoimmune or degenerative processes.

■ **Cartilage, torn** Damage to one of the semilunar cartilages of the knee, either torn away from the attaching ligament or a tear in the cartilage surface itself.

■ **Gout** Disorder of uric acid metabolism leading to uric acid crystals being deposited in the joints. Periodic attacks of acute arthritis occur.

■ **Osteoarthritis** Degeneration of the cartilage covering at the ends of the bones and the formation of an irregular surface causing limitation and pain on movement. This process is aggravated by age, obesity and excessive use.

■ **PID** Prolapsed invertebral disc. The central nucleus pulposus of the intervertebral disc ruptures through the tough fibro-cartilage ring. This protrusion may press on the adjacent nerve causing neuralgia and sciatica.

■ **Rheumatoid arthritis** Generalized disease in which the joints are damaged and may lead to ankylosis.

■ **Spondylitis** Inflammation of the joints of the spine.

■ **Spondylosis** Degenerative changes in the joints of the spine, akin to osteoarthritis.

■ **Synovitis** Inflammation of the synovial membrane of the joints.

**Cross-references**

Structure of body 13
Skeleton 20–3
Muscles 27

# Muscles

**■ Acetylcholine** Chemical produced in the transmission of nerve impulses to the muscles by the autonomic nervous system.

**■ Adrenaline** Hormone produced by the medulla of the adrenal gland. It augments the action of the sympathetic nervous system.

**■ Antagonist** In this sense one of a pair of muscles whose balance of action is such that an antagonist gradually relaxes as the other muscle contracts.

**■ Aponeurosis** Sheet of strong tissue, within the body, to which muscles may be attached.

**■ Cramp** Involuntary severe spasm of a muscle.

**■ Fascia** Thin layer of connective tissue.

**■ Fibril** Small area of contractile tissue in a muscle. Protein molecules are the basic component of fibril.

**■ Ganglion** A swelling. A nerve ganglion is where nerves join to each other outside the central nervous system.

**■ Lactic acid** One of the substances produced when muscle activity takes place.

**■ Muscle "belly"** Fleshy part of the muscle.

**■ Muscle "head"** Beginning of the muscle belly at the site of origin. A muscle may have more than one "head".

**■ Muscle "insertion"** Muscle belly contracts down to form a tendon, which is attached to either a bone or aponeurosis, called its insertion.

**■ Muscle "origin"** Usually a large area where the muscle is attached to the bone.

**■ Muscle "tone"** Constant slight contraction that keeps the fleshy part of a muscle firm.

## Skeletal muscles

### Basic facts

Muscles are contractile tissue, which can initiate or maintain movement in the body. Muscles comprise 35 to 45 per cent of the total body weight and there are over 650 skeletal muscles (those directly under the skin are shown above) controlled by the nervous system. There are three kinds of muscle (shown top right): Skeletal, under voluntary control; cardiac, only found in the heart; and smooth, as in the intestine.

### Contraction

Muscles are composed of fibrils, which shorten when chemically activated. This occurs in response to nervous stimulus. The chemical that initiates this depends on the type of muscle, but the response is the same – a shortening of protein molecules. As soon as the nerve stimulus ceases, the fibril relaxes and the muscle lengthens. Muscle tone is maintained by some of the fibrils always being stimulated to contract.

### Muscles – metabolism

When muscles contract energy is required and heat produced. This metabolism produces carbon dioxide, lactic acid, heat and water. The blood flow in to the muscles is increased to take away the metabolites, the heart rate increases and the heat has to be dissipated through the skin and by sweating. Excessive use leads to a local build-up of lactic acid as the oxygen supply is not sufficient to metabolize it. This can cause cramp.

**Types of muscles**

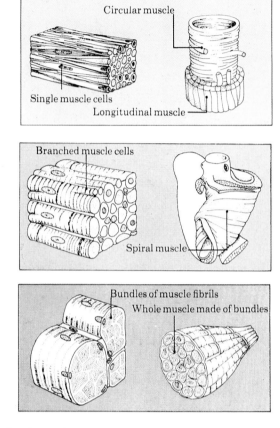

Circular muscle

Single muscle cells

Longitudinal muscle

Branched muscle cells

Spiral muscle

Bundles of muscle fibrils
Whole muscle made of bundles

## Smooth muscle

Made up of long cells arranged in bundles, each with a single nucleus. The intestine has a circular inner layer producing constriction, and an outer longitudinal layer to produce wave-like peristalsis. Controlled by the autonomic system, smooth muscle contraction is involuntary.

## Cardiac muscle

Made up of branched cells with many nuclei that interweave and form thick, spiral bands around the ventricles of the heart. These have a slow, rhythmical contraction that is increased by the cardiac pacemaker. Cardiac muscle is also controlled by the autonomic nervous system.

## Skeletal muscle

Made up of long cells, some as much as 20 centimetres in length, this type of muscle is composed of minute myofibrils and nuclei, which form bundles under the control of the voluntary nervous system. Skeletal muscle can contract and relax rapidly, but it is easily fatigued.

## Skeletal muscles – anatomy

These may be massive, like the gluteus maximus in the buttock, or minute, like the stapedius muscle inside the middle ear. Most muscles join one bone with another and have their "origin" on one immobile bone and "insertion" on the other, the one that is moved. The origin is usually on a moderately large area of bone and, with some muscles, may be from more than one area, as on the two "heads" for the biceps.

The main body of the muscle is called "the belly". The insertion is usually a tendon attached to a small area of bone, but may be an aponeurosis attached to other structures, as is the case in the back. Long tendons passing over other structures, for example at the wrist, are usually surrounded by sheaths of synovial membrane in order to reduce friction. In these areas the tendons are held down by the retinaculum.

Short tendons insert directly into the periosteum of the bone and often produce a small bony nodule on the surface of the underlying bone.

## Skeletal muscle – movement

Long muscles, like the sartorius, contract further than stronger muscles like the deltoid. This is because the arrangement of fibres is parallel, while those in the deltoid are arranged in a fan shape of small bundles inserting into the tendon.

Muscles are usually arranged in pairs (*right*), so that as one contracts (**1**) the other, the antagonist (**2**), slowly relaxes to give a smooth, controlled movement. Muscles are gently pulling against each other all the time. This is muscle tone.

When a muscle contracts and moves a joint it can only do so if the insertion is held rigid. The brain controls skeletal muscle movement and co-ordinates movement using information relayed from the muscle itself, and from the eye and the organ of balance in the ear.

Muscles requiring skilled movements, as in the hand, have one nerve to a few fibres, but in those that are used for strength, such as the gluteus maximus, one nerve will supply a large number of fibres.

**Joint movement**

1
2

# Involuntary Muscles

## Autonomic nervous system

Many of the muscles in the body are controlled by the autonomic nervous system (*right*), and are composed of sympathetic nerves (red) and parasympathetic nerves (blue). These involuntary muscles are of two kinds: Cardiac muscle, which is only found in the heart; and smooth muscle.

Smooth muscles are found all over the body – surrounding blood vessels that regulate the blood flow; in the bronchioles controlling the entry of air to the alveoli; in the iris of the eye, dilating and contracting the size of the pupil in reaction to changing light; and attached to the hairs in the skin, making them "stand on end". Larger amounts are found making up the walls of the bladder and uterus and their appendages – ureters and Fallopian tubes. They are also found about half-way down the esophagus.

**Autonomic nervous system**

Parasympathetic     **1**     Sympathetic

**Release of urine**

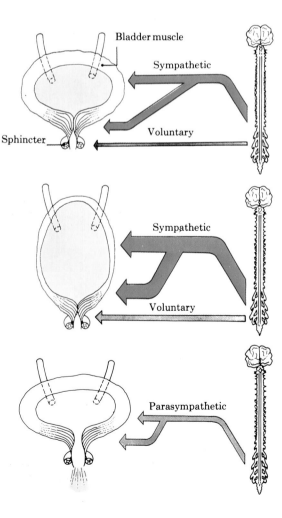

Bladder muscle

Sympathetic

Voluntary

Sphincter

Sympathetic

Voluntary

Parasympathetic

**Release of urine**
The bladder fills with urine. The sympathetic nerves to the bladder muscle make it relax and the sphincter contract. Voluntary nerves also control the sphincter so that urine can be passed at convenient times.

The urine pressure increases so the sympathetic nerve impulses intensify, keeping the sphincter firmly closed and the bladder relaxed. In the normal active male the bladder holds approximately 300 millilitres of urine.

When the bladder is full the sympathetic nerve impulses cease and the parasympathetic nerves cause the bladder to contract and the sphincter to open. Only controlled voluntary impulses can prevent the urine from being passed.

## Involuntary control

The intestinal wall is surrounded by a network of nerve fibres from both the sympathetic and parasympathetic nervous systems. The sympathetic nerves release noradrenaline in order to relax the muscle walls and contract the sphincters.

The parasympathetic nerves release acetylcholine to produce the opposite effect – contraction of the muscle walls and relaxation of the sphincters. The stretch receptors in the smooth muscle help in the control of muscle tone.

The balanced control of these two systems produces peristalsis. The autonomic nervous system (*above*) is ultimately under the control of the brain's hypothalamus. The sympathetic nerves leave the spinal cord (**1**) to reach ganglia (**2**) – nervous relay stations – which interconnect and lie on the muscles near the spine before passing on to the body.

The sympathetic nervous system prepares the body for action – dilated pupils, cold skin, raised blood pressure and pulse rate, and blood diverted from a relaxed intestine to the skeletal muscles. The parasympathetic nerves run directly from the spinal cord. The parasympathetic nervous system prepares the body for a relaxed state.

## The alimentary canal

The alimentary canal extends from the pharynx (**1**) to the rectum (**2**). When the food enter it, circular muscle fibres contract (**3**) behind it and relax in front (**4**), while the outer, longitudinal fibres (**5**) squeeze it along in a wave-like fashion. This is known as peristalsis and is the manner in which food is passed through the alimentary canal.

## Peristalsis

When food enters the mouth it is under voluntary control and can be chewed and even spat out. Once it has entered the esophagus it is no longer under conscious control and is moved down the alimentary canal by peristalsis (*above*). Peristalsis is a complex, reflex wave of contraction and relaxation. The mass of food is called a "bolus". The circular and longitudinal smooth muscle fibres behind the bolus contract and this tends to push it onwards. This tendency would be frustrated if the fibres over the front part of the bolus and those immediately in front of it did not relax and create a space into which it could move. As it moves forwards the wave of constricting muscle extends forward to squeeze the bolus from behind.

Peristaltic waves occur all the time and move the intestinal contents along the alimentary canal. Sometimes peristalsis moves backwards and forwards mixing the contents – particularly in the stomach – before it is released in small amounts at a time into the duodenum.

## Sphincters

Tight areas of circular muscles – sphincters – close off one part of the intestine from another. They only open when it is necessary for food to pass through. There are six sphincters: Esophageal sphincter, where the esophagus joins the stomach; pyloric sphincter, between the stomach and duodenum; the small sphincter, between the terminal ileum and colon; the anal sphincter, at the end of the alimentary canal, which is also under voluntary control; sphincter of Oddi, where the common bile and pancreatic ducts join the duodenum; and the sphincter in the bladder.

When food is eaten the empty stomach stimulates a giant peristaltic wave. Several of these waves pass along the entire length of the alimentary canal, moving the contents onwards. The contents of the colon are moved to the rectum, which is normally empty. The distension of the rectum produces the urge to defecate. This is why so many people do so after the first meal of the day. Antiperistaltic waves are peristalsis in reverse and may produce vomiting.

## Cardiac muscle

Cardiac muscle is arranged in thick interweaving circles around the ventricles. It has an inbuilt rhythm of contraction and relaxation which occurs about 80 times a minute. Special, electrically sensitive muscle fibres in the "pacemaker" respond to the autonomic nervous system and produce a faster, regular rhythm adjusted to the body's needs. The blood pressure is produced by the strength of the heart beat and is reflexly controlled by sensitive nerve endings responding to the pressure in the carotid arteries.

## Involuntary muscle and hormones

Adrenaline augments the action of the sympathetic nervous system; to a lesser extent thyroxine, from the thyroid gland, has a similar effect.

## Muscle and heat production

Heat is produced by peristalsis and the maintenance of muscle tone. Cooling of the body causes rapid, involuntary contractions of skeletal muscle – shivering. It is the contractions that make extra heat.

---

■ **Myasthenia gravis**
Disorder disease of the transmission of impulses between the nerve ending and the muscle, leading to fatigue and weakness when the muscles are used.

■ **Myotonia** Group of congenital muscle disorders of increased tone and stiffness.

■ **Myositis**
Inflammation of the muscle. This will commonly occur with any acute infectious illness and is a transitory affair. Rarely, there is true infection of the muscles.

■ **Polymyalgia rheumatica**
Inflammation of the muscle arteries in the elderly causing stiffness and pain of muscles and joints.

■ **Rheumatoid arthritis** Although this is a disease mainly of the joints, the muscles are also involved and will become weaker.

■ **Rupture of Achilles' tendon**
Any tendon may rupture and one of the commonest is the largest in the body, the Achilles' tendon, which may spontaneously rupture after minimal strain. This is probably due to a disorder of the collagen connective tissue of which it is made.

■ **Synovitis**
Inflammation of the synovial membranes covering the tendons and joints.

■ **Tendinitis**
Inflammation, usually due to excessive use, of the tendons.

**Cross-references**

Structure of body 13
Cell 14–15
Skeleton 20-23
Joints 24–5
Heart 32
Digestion 47
Nervous system 58

# Heart and Circulation

## The two circulations

The systemic circulation carries oxygenated blood and nutrition and returns with metabolites that have to be filtered and excreted.

The pulmonary circulation quickly exchanges carbon dioxide and oxygen through the alveolar surface in the lungs.

## The circulatory system

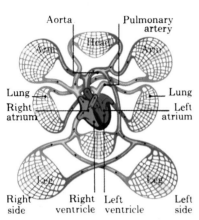

## The circulatory system

Blood is pumped into large and then smaller arteries (*above*) before reaching the arterioles and finally the capillaries, each about 1 millimetre long and one-hundredth of that in diameter. It is here that the exchange of gases takes place and interstitial fluid is formed before the blood returns into the venous system. The systemic and pulmonary circulations are based on the same system with lymphatic vessels collecting excess interstitial fluid.

## Kidney and liver circulations

Twenty per cent of the cardiac output passes through the kidneys. This blood is at high pressure to maintain adequate filtration of metabolites into the renal tubules.

The liver has a small hepatic artery and large hepatic veins due to the portal venous system. This starts as capillaries in the intestine, from the upper end of the stomach to the anus, where anastomoses with the systemic venous system occur. The blood flows in the mesenteric veins to join the splenic vein to form the hepatic portal vein and enter the liver, where it again forms capillaries to enter the hepatic sinusoids.

## Anatomy of the circulation

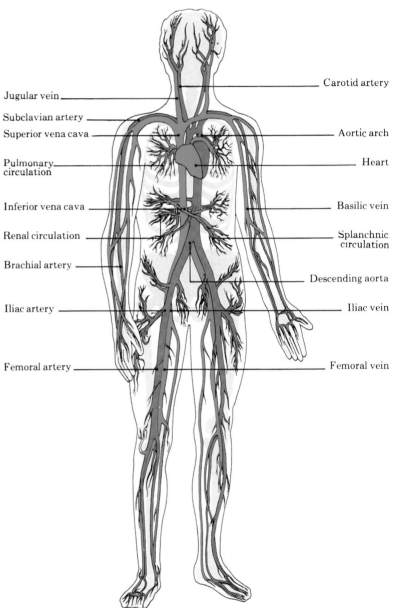

## Anatomy of the circulation

The blood is carried in a 96,500 kilometre network of blood vessels (*above*). The aorta leaves the aortic valve where its first two branches, the coronary arteries, subdivide into brachial and carotid arteries. The aorta then runs in front of the spine and behind the esophagus to the abdomen, with small and large branches to the vertebrae, diaphragm and intercostal muscles. In the abdomen it divides just above the pelvis into the common iliac arches to the legs.

There are four main branches: Gastric to the stomach; splanchnic to the intestine; renal to the kidneys; and splenic to the spleen. The venous blood returns to the superior and inferior venae cavae and thus to the right atrium.

## The heart in health

A baby's heart rate is about 130 beats a minute; this falls to around 70 in the resting adult. A trained athlete can raise the rate to about 200 with great exertion. The heart rate is controlled from the cardiac centre in the medulla; parasympathetic fibres in the vagus slow the rate and sympathetic nerves and adrenaline increase it in response to stressful situations, such as emotional disturbance and fear.

Chemoreceptors, in the carotid body, are sensitive to low oxygen levels and cause a tachycardia. Baroreceptors in the carotid sinus reflexly control the blood pressure.

## The cerebral circulation

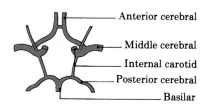

- Anterior cerebral
- Middle cerebral
- Internal carotid
- Posterior cerebral
- Basilar

## The cerebral circulation

The two vertebral arteries join to make the basilar artery, which communicates with the two internal carotid arteries to make a circle of vessels (*above*) from which the anterior, middle and posterior cerebral arteries supply the cerebral hemispheres. This ensures an adequate blood supply at all times.

## Blood pressure

Blood pressure has two measurement points: Systolic, the highest pressure reached; the diastolic, the lowest, when the aortic valve is closed and the left ventricle is in diastole. The elasticity of the artery wall allows expansion and then maintains arterial blood flow, during diastole, by contraction. A loss of elasticity with ageing is associated with higher systolic pressures. Blood pressure is measured with a mercury column in a sphygmomanometer. A relaxed, healthy adult has a blood pressure of around 120 millimetres systolic, and 70 millimetres diastolic, which is written as 120/70.

Blood pressure rises slightly on standing as there is a nervous contraction of blood vessels in the legs.

## A heart beat

**A**

**B**

## A heart beat

In atrial diastole (**A**) blood flows from the inferior and superior venae cavae into the right atrium (**1**) and from the four pulmonary veins into the left atrium (**2**). The flow is increased during inspiration as the negative intrathoracic pressure also "sucks" blood into the heart as well as air into the lungs. This results in sinus arrhythmia.

When ventricular systole ceases (**B**) the intraventricular pressure drops and the two atrioventricular valves – the tricuspid (**3**) and the mitral (**4**) – float open and blood starts to flow from the atria (**1, 2**) into the ventricles. The sino-atrial node then initiates atrial systole and the blood is pumped past the valves, which are fully open, into the dilated ventricles.

**C**

**D**

Atrial systole ceases (**C**) when the electrical impulse reaches the atrioventricular node and passes down the Bundle of His to start ventricular systole. Once atrioventricular valves (**3, 4**) snap closed, the chordae tendineae and papillary muscles prevent them bursting back into the atria. Venous blood can again flow into the atria (**1, 2**), during atrial diastole and ventricular systole.

The rapidly increasing ventricular pressure (**D**) throws open the aortic (**5**) and pulmonary (**6**) valves with blood streaming into the systemic and pulmonary circulations. The elasticity of the arterial walls causes the valves (**5, 6**) to snap closed at the end of ventricular systole. The "snap" of the opening and closing heart valves can be heard as "lub-dub" through the chest wall.

- **Chemoreceptor** Nerve ending that is sensitive to the slightest chemical change.
- **Chordae tendineae** Fibrous strands from the papillary muscles to the edges of the atrioventricular valves.
- **Coronary** "Crown" of two blood vessels supplying the blood to the heart muscle.
- **Diastole** Period of relaxation of the heart.
- **Ductus arteriosus** Foetal blood vessel that connects the aorta and pulmonary artery and which closes at birth.
- **Endocardium** Layer of smooth cells lining the internal surface of the heart.
- **Endothelium** Smooth layer of cells lining the internal surface of the blood.
- **Foramen ovale** Inter-atrial opening that closes at birth.
- **Iliac artery** Aorta divides into two iliac arteries.
- **Lymphatic vessel** Thin-walled vessel that collects interstitial fluid and returns it, as lymph, to the venous system located in the neck.
- **Medulla oblongata** Part of the brain stem
- **Mesenteric** Supplying the intestine
- **Mitral valve** Valve between left atrium and ventricle
- **Myocardium** Muscle of the heart.
- **Papillary muscle** Ventricular muscle to which the chordae tendinae are attached.
- **Pericardium** Sac surrounding the heart.
- **Portal system** Vein beginning and ending in capillaries.
- **Pulmonary** Refers to structures associated with the lungs.
- **Pulmonary valve** Valve between the right ventricle and pulmonary artery.

# The Heart and Blood Vessels

## Position

The heart (*right*) lies in the thorax behind the sternum and in front of the descending aorta and esophagus. It rests on the central ligament of the diaphragm muscle. On either side are the two lungs. Above are the main blood vessels and the bifurcation of the trachea into the two main bronchi.

## Basic facts

The heart weighs about 300 grammes and is about the size of a grapefruit. It has two atria, two ventricles and four valves. It receives blood from the two venae cavae and four pulmonary veins and expels it into the aorta and pulmonary artery. It pumps 9,000 litres of blood a day at a rate varying between 60 and 160 beats a minute.

## Anatomy

The heart is surrounded by the fibrous pericardium containing the serous pericardial sac holding a small amount of fluid that allows friction-less movement. It consists of two pairs of chambers – atrium and ventricle – that act as separate pumps. The right side pumps deoxygenated blood through the lungs – the pulmonary circulation. The left side pumps oxygenated blood from the lungs – the systemic circulation.

## Atria and ventricles

The blood from the superior and inferior venae cavae enters the right atrium. The four pulmonary veins take blood to the left atrium.

The atrioventricular valves, tricuspid in the right and mitral in the left, have special muscles – the papillary muscles – and fine tendons – chordae tendineae – attached to the edges of the valve cusps to prevent them rupturing back into atria during ventricular systole.

The left ventricle has thicker muscle than the right to sustain a higher systemic blood pressure. The ventricles are closed by the aortic and pulmonary valves.

The heart is lined with endocardium, and divided into halves by the interatrial and interventricular septa.

The remnants of the foetal circulation can still be seen in the fibrous band of the ductus arteriosus.

**The heart**

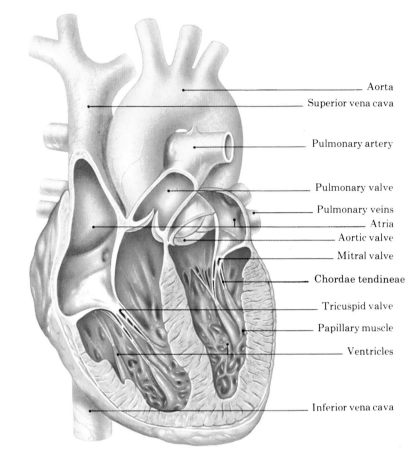

- Aorta
- Superior vena cava
- Pulmonary artery
- Pulmonary valve
- Pulmonary veins
- Atria
- Aortic valve
- Mitral valve
- Chordae tendineae
- Tricuspid valve
- Papillary muscle
- Ventricles
- Inferior vena cava

**Electrical system**

Sino-atrial node

Atrio-ventricular node

Bundle of His

**Coronary circulation**

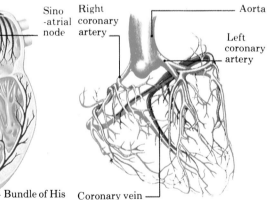

Sino-atrial node — Right coronary artery

Aorta

Left coronary artery

Coronary vein

## The electrical system

To keep the heart (*above*) beating the sino-atrial node, "pace-maker", in the right atrium sends impulses through the two atria, causing atrial systole. It then stimulates the atrioventricular node to pass rapidly down the Bundle of His to cause ventricular systole.

## Coronary circulation

The myocardium has its own blood supply (*above*) from the left and right coronary arteries – the first branches of the aorta – supplying different areas of muscle with a few anastomoses between them. The venous blood is drained into the right atrium by the coronary veins.

# The blood vessels

The arteries carry blood away from the heart and the veins return it. Both arteries and veins (*right*) are made in the same way with four layers: A protective fibrous coat; a middle layer of smooth muscle and elastic tissue, which is thickest in the largest arteries; a thin layer of connective tissue; and a smooth layer of cells – endothelium.

## Arteries

The blood in the arteries is under high pressure, greater in the systemic than the pulmonary circulation, so the arterial walls are stronger, with thicker muscular and elastic layers. This allows the artery to expand with the surge of pressure at each heart beat and thus smooth out the blood flow. This can be felt as a "pulse" in arteries near the surface.

The larger arteries divide into smaller ones and finally into arterioles, where the blood flow can be controlled by autonomic nerves supplying the smooth muscle. This allows adjustments of blood flow to various areas when required.

## Veins and venous blood flow

The blood reaches the venous system through the minute capillary vessels. It is through the capillary wall that oxygen and carbon dioxide, food and metabolites are exchanged with the interstitial fluid. Most of the interstitial fluid returns to the venous system, but some is collected by the lymphatic vessels.

The returning venous blood moves slowly due to low pressure and the veins can collapse or expand to accommodate variations in blood flow (*right*). Movement relies on the surrounding muscles, which contract (**1**) and compress the vein. Pulsation of adjacent arteries (**2**) has a regular pumping effect.

Semilunar valves (**3**) are found at regular intervals throughout the larger veins and these allow the blood to move only in one direction. They are commoner in the legs.

The veins frequently anastomose with each other so that the blood flow can alter direction if there is any constriction or pressure from movement of muscles or ligaments.

## The blood vessels

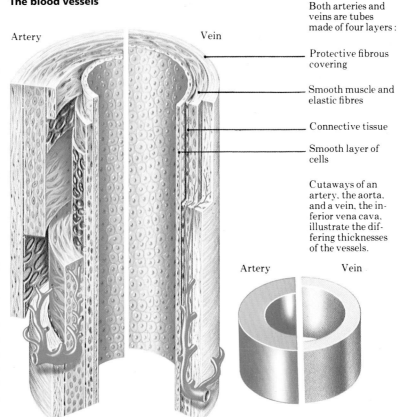

Artery

Vein

Both arteries and veins are tubes made of four layers :

Protective fibrous covering

Smooth muscle and elastic fibres

Connective tissue

Smooth layer of cells

Cutaways of an artery, the aorta, and a vein, the inferior vena cava, illustrate the differing thicknesses of the vessels.

Artery  Vein

## Venous blood flow

■ **Myocarditis**
Inflammation of the heart muscle, such as rheumatic myocarditis.

■ **Pericarditis**
Inflammation of the pericardium, such as tuberculous pericarditis.

■ **Rheumatic fever**
Raised temperature and inflammation of connective tissues – occurring mostly among children.

■ **Septal defect**
"Hole in the heart".

**Diseases of blood vessels**

■ **Arteries** – Arteriosclerosis Narrowing of the vessels leading to blockage and death of the area supplied, e.g. coronary or cerebral thrombosis.

■ **Hypertension**
Raised blood pressure, usually for no obvious reason.

■ **Veins** – Varicose Congenital absence of valves leading to distortion of veins by the pressure.

**Electrical irregularities**

■ **Atrial fibrillation**
Irregular, rapid heart beat due to disorder of the sino-atrial node.

■ **Heart block**
Failure of the electrical impulse to reach the ventricles, which then contract at their "natural" rhythm of 40 beats every minute.

**Cross-references**

Muscles 26, 27, 29
Respiration 38–43

# Blood

## Blood formation

**Blood formation**

### Definitions

■ **Adenoid** Pad of lymphoid tissue on the posterior wall of the posterior nasal cavity.

■ **Albumen** Simple plasma protein.

■ **Antibody** Specific substance produced by the body in an attempt to dispose of a foreign substance.

■ **Antigen** Any substance introduced in the body that causes the formation of an antibody.

■ **Basophil** Blood cell that is stained by basic staining dye.

■ **Blood group** Classification of different types of blood by their antigen characteristics.

■ **Carbamino-haemoglobin** Haemoglobin combined with carbon dioxide.

■ **Carbonic anhydrase** Enzyme that causes water to combine with carbon dioxide to form carbonic acid.

■ **Eosinophil** Blood cell stained red.

■ **Fibrin** Protein that causes strands in clotting.

■ **Fibrinogen** Soluble precursor of fibrin in the circulation.

■ **Folic acid** Member of the vitamin B group.

■ **Gamma globulin** Particular group of plasma protein associated with immunity.

■ **Globulin** Major group of plasma proteins.

■ **Haemoglobin** Complex protein that gives the colour to red cells and is essential for the carriage of oxygen. It contains the element iron.

■ **Haemolysis** Breakdown of red blood corpuscles releasing haemoglobin.

■ **Krupffer's cells** Modified macrophage cells of the reticulo-endothelial system in the liver's sinusoid.

Labels on figure: Liver — Spleen — Lymph glands — Stem cell — Red blood cell — Thrombocytes or platelets — Lymphocyte — Monocyte — Bosophil white blood cell — Eosinophil white blood cell — Neutrophil white blood cell — Blood-forming marrow — Nutrient artery

## Blood

Blood is the fluid that is pumped round the body in the circulatory system. It is made up of many components and has many functions. A baby has about 250 millilitres, a man over 5 litres and a woman rather less, about 4 litres of blood. It consists of a yellow fluid – the plasma – in which red and white blood corpuscles and platelets are suspended.

The capillaries allow fluid to escape from the blood. The cells and large proteins are left in the vessel and this fluid can now become the interstitial fluid. This will either return to the capillary or join the lymphatic system. In the brain it helps form the cerebro-spinal fluid and in the eyes the intraocular fluid. Blood amounts to about one-third of the total interstitial fluid.

## Blood formation

Red and white blood cells (*above*) are formed in the bone marrow – the dotted areas of bone. In addition, white cells are produced by the liver, spleen and lymph nodes. The stem cells in the bone marrow produce the red and white blood corpuscles (magnified 1,000 times). The red corpuscles enter blood vessels through the nutrient circulation. The white corpuscles – the polymorphonucleocytes, basophils, neutrophils and eosinophils – have segmented nuclei. Monocytes and lymphycytes have single nuclei. Thrombocytes – platelets – are also formed from the stem cell system by fragmentation and are involved in blood clotting.

In infants, all the bones make red blood cells, and before birth the liver and spleen also produce them.

## The plasma

The plasma occupies 55 per cent of the blood volume. It is 90 per cent water, 7 per cent proteins and lipoproteins with the remaining 3 per cent made up of small molecules – salts, glucose, vitamins, hormones, urea, amino-acids and dissolved carbon dioxide.

The main proteins are albumen, globulins and fibrinogen. They help to maintain the osmotic pressure to hold fluid in the circulation and counteract the blood pressure that forces it out of the capillaries. The proteins also repair the body, the gamma-globulins defend it against infection, and fibrinogen may break down to form fibrin when clotting occurs. Serum is plasma without fibrinogen and is seen as the thin fluid oozing from grazed skin.

## Constituents of the blood

Polymorphonuclearcytes immature to mature form

Plasma

White blood cells

Red blood cells

Small and large lymphocyte    Monocyte

Platelets

Red blood cells

### Red blood corpuscles

Each red corpuscle (*above*) measures about 7.5 microns (thousandths of a millimetre) in diameter and is biconcave in shape. It contains haemoglobin, which gives it a red colour. There are 5 to 6 million red corpuscles per cubic millimetre of blood.

Red corpuscle production requires vitamins of the B complex – principally vitamin $B_{12}$ – and folic acid; vitamin C; and, as well as proteins, the elements iron, copper and cobalt. The corpuscles are formed from the marrow stem cells, but less than 1 per cent of the circulating red cells have nuclei as the nucleus is removed when it reaches maturity.

The red corpuscle survives about 120 days and the damaged and old cells are removed by the reticulo endothelial system in the spleen and liver. The red corpuscles have only one function: The transport of oxygen by combining with haemoglobin. Healthy men have about 15 grammes of haemoglobin in 100 millilitres of blood and women 13 grammes.

### White blood corpuscles

There are normally about 8,000 white corpuscles (*left*) in every cubic millilitre of blood. There are two main varieties: Polymorphonuclear leucocyte (polymorphs) and lymphocyte. Polymorphs are so called because of their nuclei. Formed in the bone marrow stem cells, they make up two-thirds of white cells. The immature cell has one nucleus, but as it matures a string of up to five nuclei forms. However, they survive for less than a week.

There are three different kinds of polymorphs, each can be distinguished by its microscopic staining character: Eosinophils, basophils and neutrophils.

Polymorphs act by phagocytosis; they have two functions: Destruction of invading bacteria; and removal of dead and damaged tissue. If the polymorphs are killed pus is formed. The eosinophils react in allergic states.

Lymphocytes are formed in the lymph glands, spleen and thymus. They help with the immunity of the body. The monocytes are larger cells, distinct from lymphocytes, and function mainly by phagocytosis.

Platelets (thrombocytes) are tiny particles formed in the bone marrow; there are about 250,000 per cubic millimetre lasting less than a week. They are concerned with clotting.

■**Lipo-protein** Substance formed from both fat and protein.

■**Lymph** Fluid collected by the lymphatic vessels from the interstitial spaces.

■**Lymphatic system** Vessels that carry lymph and return it to the venous system.

■**Lymphocyte** White blood cell with clear cytoplasm and a single nucleus formed in lymphoid tissue.

■**Lymphoid tissue** Specialized tissue that produces lymphocytes, filters out foreign substances and helps in developing the body's immune system.

■**Macrophage** Fixed tissue cell that is involved in the phagocytosis of bacteria and foreign substances.

■**Marrow** Blood-forming tissue in the centre of bone.

■**Monocyte** Circulating phagocytic cell.

■**Neutrophil** Body cell that does not stain with basic or eosinophilic dye.

■**Pus** Collection of dead white blood cells.

## Blood clotting

Tissue factor
Fibrinogen
Platelet

Plasma factor

A          B          C          D

Red blood cell    Fibrin

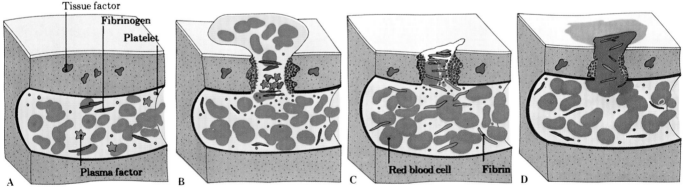

### Blood clotting

Blood clotting prevents excessive blood loss from a wound. Normally, circulating blood contains red cells, platelets, plasma, clotting factors and fibrinogen. Tissue-clotting factors lie trapped within cells surrounding each blood vessel (**A**). When damage occurs blood escapes from the broken vessel. Platelets congregate at the site and help plug the wound. Tissue-clotting factors are released (**B**). The reaction of the platelets with plasma and tissue-clotting factors converts the soluble fibrinogen into insoluble threads of fibrin. The fibrin forms a mesh across the break (**C**). Platelets and blood cells become trapped in the mesh. The jelly-like mass shrinks and serum oozes out, leaving a clot (**D**).

# Blood Groups and Immunity

## Blood groups

Although the red cells in different people look the same they are, in fact, dissimilar. They can be divided into four main groups, A, B, AB, and O (*below*). The surface of the cells in each group is different and will act as an antigen to plasma from another group, which carries the antibody. This causes the cells to stick to each other by the process of agglutination. An individual with Group A cells will carry the antibody B in his plasma, those in group AB do not carry either antibody, while those in group O have both antibodies but the cells do not have either antigen. Thus blood from any group can be transferred into those of group AB, as they do not carry antibodies; group AB individuals are known as "universal recipients". A group O recipient can only receive blood from another group O donor, but can give blood to anyone, and thus is known as a "universal donor".

Antibodies in the plasma of the donor blood are quickly diluted by the recipient and the concentration is therefore too low to cause agglutination. The two commonest groups in western Europe are Groups A and O, each found in about 45 per cent of the population, group B in about 10 per cent and group AB is found in less than 5 per cent.

The antigen-antibody reaction not only causes agglutination, but also haemolysis – the breakdown of the red blood cells releasing haemoglobin into the circulation. This is an "incompatible transfusion reaction" and can lead to fever, jaundice, kidney failure from blocking of the tubules with haemoglobin, and even, in some cases, death.

## Rhesus factor

In addition to the four main blood groups there are many other minor ones: Rhesus, NNS, P, Kell Lewis, Duffy, Lutheran, to name a few. The most important is the Rhesus factor, named after the Rhesus monkey, in which it was first discovered. The presence or absence of this antigen makes the individual either Rhesus positive or Rhesus negative. Antibodies are not found in Rhesus negative people unless they have been transfused with Rhesus positive blood. About 15 per cent of the population is Rhesus negative.

It is therefore possible to be Group A Rhesus positive or negative and this is usually written A Rh +ve or A Rh -ve. A true universal donor has to be O Rh -ve. Blood from the recipient and donor are cross-matched before use to check that agglutination does not occur.

## Pregnancy and the Rhesus factor

Particular problems arise in Rhesus negative pregnant mothers with Rhesus positive fathers. In the last weeks of pregnancy small numbers of the baby's red cells escape through the placenta into the mother's circulation. This will not cause trouble in the first pregnancy but in subsequent pregnancies agglutination reactions may occur.

The mother's Rhesus antibodies will invade the baby's circulation, destroying the red cells and leading to anemia; this necessitates an exchange transfusion for the baby at birth with Rhesus negative blood. Provided the possibility is recognized early enough anti-Rhesus immune globulin injections can be given to the mother, to remove Rhesus positive cells from her blood before she develops her own anti-Rhesus globulin antibody.

**Blood: Oxygen and carbon dioxide**

**Blood: Oxygen and carbon dioxide**
Oxygen combines with haemoglobin (**1**) in the red blood corpuscles (*above*) to form oxyhaemoglobin. It would require 75 times as much blood to carry the oxygen if haemaglobin did not exist.

Carbon dioxide (**2**), from the interstitial fluid, dissolves in the plasma and red cells, where the enzyme – carbonic anhydrase – forms carbonic acid. A little also combines with haemoglobin to form carbamino-haemoglobin. The acidity of carbonic acid forces oxygen out of oxyhaemoglobin (**3**) into the interstitial fluid with a formation of sodium and potassium bicarbonate. A reverse diffusion occurs in the capillaries of the lung alveoli with loss of carbon dioxide (**4**) and reoxygenation of the haemoglobin (**1**).

## Blood groups

| Receptor Donor | A | B | AB | O |
|---|---|---|---|---|
| A | | | | |
| B | | | | |
| AB | | | | |
| O | | | | |

## Immunity and resistance to infection

The body's first line of defence – the skin – is a barrier against infection; it is a waterproof layer incorporating bactericidal sweat and sebaceous secretions. Internal surfaces are covered with a layer of mucus; the mucus is produced as a result of the cleansing effect of the ciliate cells lining the sinuses and bronchi and the antiseptic hydrochloric acid in the stomach. Peyer's patches in the small intestine engulf the remaining bacteria. The adenoids, tonsils and lingual pad of lymphoid tissue filter off bacteria in the nasopharynx. When infection enters the body there is an immediate response by the white blood cells, which invade the area and form pus if sufficient numbers are killed.

Throughout the body the reticulo-endothelial system of macrophage cells will engulf foreign particles and bacteria. These cells are found principally within the bone marrow, lymphatic system and spleen. In the liver they line the sinusoids and are known as Kupffer's cells.

Circulating plasma globulins give protection against any infection, but this is greatly increased by successful resistance to an infection. This immunity follows the production of specific antibodies by the lymphoid tissue and the reticulo-endothelial system. The thymus gland produces a hormone to stimulate immunity by encouraging the production of both lymphoid tissue and lymphocytes. After adolescence, however, it ceases to function.

Active, often permanent immunity is given by natural infection, injections of killed organisms, for example, diphtheria, or mildly infective living viruses, such as measles or, by mouth, poliomyelitis. Passive, temporary immunity is given by injections of gamma globulin prepared from a person who has been actively immunized, for example, human antitetanus globulin.

To achieve the maximum effect some vaccines have to be given in two or three doses, and a "booster" dose may be needed after a time to sustain immunity.

## The lymphatic system

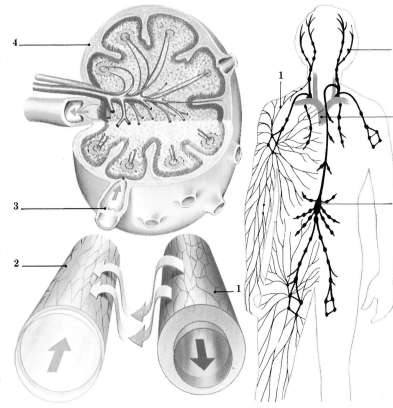

## Lymphatic system

The lymphatic system (*above*) is a network of vessels that drains the interstitial fluid back into the blood circulation. The fluid passes out of the capillary (**1**) and either into the vein or into the smallest, thin-walled lymph vessel (**2**). These lymphatic vessels join together to form large channels and finally reach the thoracic duct running next to the descending aorta. This duct joins one of the main branches of the superior vena cava (**5**).

Valves (**3**) keep lymph flowing in one direction. Lymph glands – nodes (**4**) – are found throughout the body and at places where lymph vessels unite, in the groin, axillae and neck and adjacent to branches of the aorta and inferior vena cava (**6**). They have three main functions: To filter out and destroy foreign substances; to produce lymphatic cells; and to produce antibodies. Lymphoid tissue – specialized lymph glands – are the tonsils, adenoids and Peyer's patches in the ileum.

## The spleen

The spleen (*left*), in the left upper abdomen, is a mass of lymphoid tissue. The white pulp produces lymphocytes and is scattered throughout the red pulp, in which macrophages filter out old or damaged cells, debris and bacteria. The spleen is also a store of iron from the broken red cells; and, like other lymphoid tissues, is an aid in the development of immunity. Removal of the spleen may have adverse effects, particularly in children.

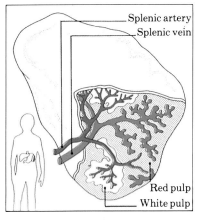

Splenic artery
Splenic vein
Red pulp
White pulp

### Common diseases

■ **Addisonian anaemia** Larger than usual red blood cells that are fewer in number due to a deficiency of vitamin B12 and associated with the destruction of the super-renal glands.

■ **AIDS (Acquired Immune Deficiency Syndrome)** Caused by the HIV (Human Immune deficiency Virus) which prevents the white blood cells reacting to even the mildest of infections.

■ **Anaemia** Lowered number of red blood cells in the circulation leading to a reduced ability to carry oxygen to the tissues.

■ **Coagulation defects** These lead to spontaneous bleeding. This may be due to an absence of platelets.

■ **Haemophilia** Sex-linked coagulation defect due to the congenital absence of a necessary protein.

■ **HIV – Human immunodeficiency viruses** (HIV-1 and HIV-2) infect T-cells and depress immunity allowing unusual diseases to occur – acquired immune deficiency syndrome, (AIDS).

■ **Leukaemia** Cancer of the white blood cells. This may be rapidly progressive – acute; or very slow – chronic.

■ **Myeloma** Type of malignant disease arising in the bone marrow and causing disorders of both the red and white blood corpuscles. It is a chronic disorder.

■ **Thrombocytopenia** Absence or severe reduction in the platelets, leading to spontaneous or post traumatic bleeding.

### Cross-references

Heart 30–3
Respiration 38–43
Pregnancy 80–5

# Respiration

**The lungs**

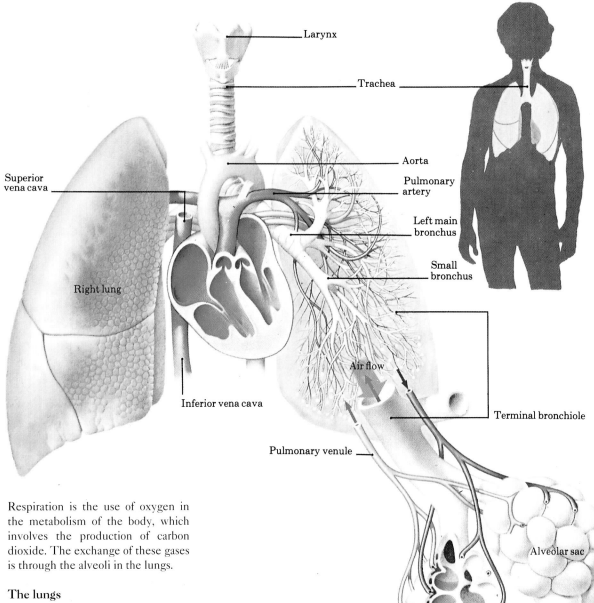

Respiration is the use of oxygen in the metabolism of the body, which involves the production of carbon dioxide. The exchange of these gases is through the alveoli in the lungs.

### The lungs

The two lungs are housed (*above right*) in the protective cage formed by the ribs radiating from the spine at the back to the sternum at the front. The curved diaphragm muscle spans the base of this beehive-shaped cage and is pierced by the oesophagus, aorta and inferior vena cava.

The lungs (*above*) are separated by the heart, blood vessels and oesophagus. The top – apex – of each lung reaches behind the first rib to the level of the clavicle, with the base resting on the diaphragm.

The left lung, consisting of two lobes, is slightly smaller than the right, with three lobes.

Each lung is surrounded by pleura. This is a double layer of membrane lining the inside of the ribs and completely enveloping the lung; it contains some lubricating fluid.

Air enters the body through the nose and mouth to reach the back of the throat – pharynx – before passing through the voice-box – larynx – to reach the windpipe – trachea. The trachea is about 12 centimetres long. It is kept open by C-shaped carti-

lages, which, when the trachea divides, continue into the two main bronchi to the lungs.

The bronchi subdivide as smaller bronchi to the lobes of each lung. Each lobe is composed of segments supplied by one bronchus. Bronchioles subdivide into the alveolar ducts of the alveolar sacs containing the individual alveoli. The pulmonary blood vessels, lymph ducts and nerves enter the "root of the lung" with the bronchi.

## Breathing

The lungs (*opposite page*) occupy most of the thoracic cavity (**A**). They are elastic structures and their elasticity helps with the breathing movements.

The main muscles for breathing are the diaphragm and the intercostal muscles, which run between the ribs. The accessory muscles that can sometimes be used are the shoulder, neck and abdominal muscles. They are used to help regain breath after running, by resting the arms on a table. This rigidity of the arms allows the arm muscles to move the chest wall instead of the chest wall acting as the firm structure for arm movement to aid in breathing.

During normal breathing the diaphragm does most of the work. It contracts, becomes flatter, and the rib cage expands (**B**). This increases the volume of the thorax and air is drawn down the trachea into the lungs – known as inspiration. Expiration takes place passively by the natural elasticity of the lung tissue and is effortless movement (**C**).

More forceful active expiration can be produced by using the intercostal and abdominal muscles. The intercostal muscles can help the diaphragm with inspiration.

The resting respiratory rate in a healthy adult is 12 times a minute; a baby breathes at twice this speed.

A normal breath is 500 millilitres of tidal air, but only 350 millilitres is fresh air, as the first air entering the lungs is 150 millilitres of dead air space – the volume of air in the nose, trachea and bronchi. The total amount of air – the vital capacity – that can be inhaled is increased by forced inspiration and expiration. There is always some air left in the lungs – the residual volume.

| | |
|---|---:|
| Vital capacity | 4,000 ml |
| Tidal air | 500 ml |
| Expiratory reserve | 1,000 ml |
| Inspiratory reserve | 2,500 ml |
| **Residual volume** | **1,000 ml** |
| Total lung capacity | 5,000 ml |

## Breathing

**A** At rest  **B** Inspiration  **C** Expiration

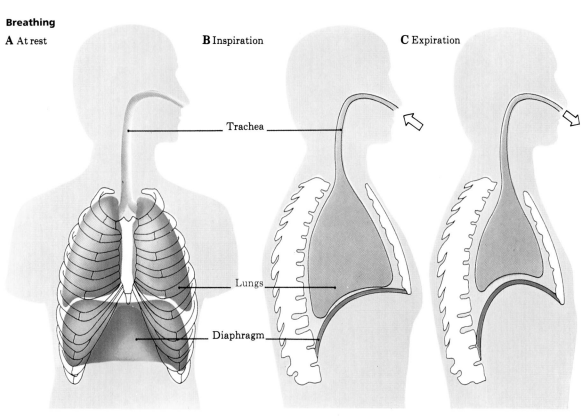

Trachea

Lungs

Diaphragm

## The nose, pharynx and the lungs

As air enters the nose larger dust particles are filtered out by the hairs in the nostrils. The air passes through the nose, where there is a large area of moist mucus membrane. This helps to humidify and warm it to body temperature. On cold, dry days the air is insufficiently warmed and moistened by the nose, so it will make the throat slightly sore – an effect often felt down to the trachea.

When air reaches the pharynx it passes through a region of lymphoid tissue in a ring around the back of the nose and throat. This consists of the adenoids on the posterior wall of the back of the nose, the two tonsils located at the sides of the back of the mouth and an adenoid-like pad of tissue found on the pharyngeal part of the tongue. This lymphoid tissue removes bacteria and viruses from air that has been inhaled.

Air enters the lungs, where the smaller particles of dust are collected in the bronchioles. The lining of the bronchi and bronchioles is moist and covered with cilia that massage the mucus and debris upwards to the trachea and finally, over the larynx, into the oesophagus. Any dust particles in the alveoli are removed by the macrophage cells – the large white blood cells present in the walls of the alveolar sac.

■ **Expiratory reserve** Additional amount of air that can be breathed out after the tidal air has been expired. This amounts to 1000 ml.

■ **Glottis** Opening between the vocal cords with the part of the larynx concerned with voice production.

■ **Haemoglobin** Complex protein contained in the red blood cells that combines with oxygen.

■ **Inspiration** Breathing air into the lungs.

■ **Inspiratory reserve** Amount of extra air that can be breathed into the lungs after the normal tidal air. This amounts to 2500 ml.

■ **Intercostal muscle** Two layers of muscles that run between each rib bone.

■ **Intercostal nerve** Nerve that runs from the spinal cord to supply the intercostal muscles.

■ **Larynx** The Adam's apple, a combination of cartilage and muscle lined with mucous membrane at the top of the trachea. It contains the glottis, muscles and posterior laryngeal cartilage.

■ **Lobe of lung** The right lung is divided into three sections, or lobes, and the left into two lobes.

■ **Lung** One of a pair of the respiratory organs occupying the major part of the thoracic cavity. It is concerned with the exchange of oxygen and carbon dioxide.

■ **Macrophage cell** Large cell derived from the blood system that is found in the connective tissues. It "swallows" foreign substances as well as bacteria.

■ **Oesophagus** Muscular tube, lined with mucous membrane, which runs from the pharynx to the stomach. It is the most posterior structure in the thoracic cavity.

# Respiration and Metabolism

### Anatomy of the alveolar sac

The anatomy of these minute grape-like alveoli is very delicate (*right*). They are held open by a framework of strands of fibrous connective tissue through which the terminal bronchioles, only 0.3 millimetres in diameter, pass and open into the individual alveolus.

The alveoli are lined by a thin layer of cells that coats the fibrous strands and the surrounding mesh of blood capillaries. The alveoli are kept moist by a thin film of water that is essential in the diffusion of oxygen and carbon dioxide in and out of the capillaries. Any dust, soot or bacteria that reaches them is ingested by the defensive macrophages in the moisture on the inner surface of the lining. Any excess fluid in the alveolar walls is drained away by the pulmonary lymphatic vessels.

The alveoli of an adult have a total surface area of 70 square metres; the whole breathing apparatus is designed to bring fresh air as close as possible to the blood.

### Respiration and metabolism

The metabolism of the body requires oxygen to burn food, which is then converted into energy, carbon dioxide and waste substances. The oxygen is obtained from the circulating oxyhaemoglobin in the red blood corpuscles. There is only a small amount of oxygen dissolved in the plasma. The carbon dioxide is entirely dissolved in the plasma. Exercise or any fever increases the breakdown of food. The increase in the production of carbon dioxide creates the need for more oxygen, leading to an increased respiratory rate.

Some diseases, such as diabetes, may cause an increase of acid production, not carbon dioxide. This acid acts as a stimulant to cause an increased respiratory rate. The body is unable to differentiate between the acidity of carbon dioxide and acid due to other causes.

Diseases that increase the metabolic rate, such as over-activity of the thyroid gland – hyperthyroidism – contrast with the physical and respiratory sluggishness of an underactive thyroid – myxedema.

**The alveolar sac**
magnified about 500 times

Alveolar walls composed of a single layer of cells, capillaries and connective tissue.

Connective tissue fibres form a frame around the alveoli and support their walls.

Alveolar lining is moist, essential for the transmission of carbon dioxide and oxygen.

Pulmonary venules carry oxygenated blood back to the left atrium of the heart.

Terminal bronchiole, lined with ciliated cells to remove dust, leading to alveolar sacs.

Lymph vessels carry excess interstitial fluid away from the alveolar sac taking it to the thoracic duct.

Pulmonary arteriole containing deoxygenated blood.

Macrophage on the inner alveolar surface.

Blood capillaries form a meshwork around the alveolus allowing the exchange of gases.

■ **Oxyhaemoglobin** Condition of haemoglobin when oxygen is combined with it.

■ **Pharynx** Cavity between the mouth and the posterior nasal cavity, above, and the beginning of the oesophagus.

■ **Phrenic nerve** One of a pair of nerves, arising in the neck, supplying the diaphragm.

■ **Pleura** Sac-like structures that lines the inner surface of the thoracic cage and surrounds each lung.

■ **Post-nasal cavity** Dilatation of the back of the nasal passages, above the pharynx into which the Eustachian tubes open.

■ **Pulmonary** Refers to structures associated with the lungs.

■ **Residual volume** Air that remains in the lungs after forced expiration, about 1000 ml.

■ **Segment of lung** Each lobe is subdivided into smaller sections, segments.

■ **Sinuses** Cavities within the skull that are lined with mucous membrane. They lighten the weight of the skull and act as resonant chambers for production of speech.

■ **Speech centre** Area of the brain concerned with the production of speech.

■ **Tidal air** Amount of air that is normally breathed in and out during quiet respiration, about 500 ml.

■ **Tonsil** One of a pair of lymphoid structures found at the back of the mouth.

■ **Trachea** Windpipe from the larynx to bifurcation of bronchi, about 12 cm long and kept open by C-shaped cartilages.

■ **Vital capacity** Total amount of air that can be breathed in and out of the lungs, 4,000 ml.

# The Lung – Oxygen and CO$_2$

**Exchange of gases in alveolus**

O$_2$ in blood    CO$_2$ in blood

O$_2$ dissolved in moist film
Air rich in O$_2$
Air rich in CO$_2$
CO$_2$ dissolved in moist film
Pulmonary venule
Pulmonary arteriole
Alveolus
Capillary

## The exchange of gases

The exchange of gases (*above*) has less than a second to take place. The carbon dioxide has to pass out of the plasma, where it is dissolved, through the capillary wall into the minute space between the capillary and the alveolar walls. It then passes through the alveolar walls into the thin film of moisture that lines each alveolus. Both carbon dioxide and oxygen have to dissolve in this moist layer on their way out of and into the blood. The gases move by diffusion – movement from an area of high to low pressure. The oxygen passes in the opposite direction to the carbon dioxide and combines with the haemoglobin in the red blood corpuscles to form oxyhaemoglobin. The capillaries are so narrow that only one red blood cell at a time can get through. The oxygenated blood returns to the left atrium through the pulmonary venous system, which drains alongside the bronchioles and bronchi.

Inspired air contains 20 per cent oxygen, 0.03 per cent carbon dioxide and the rest is made up of nitrogen. Expired air contains 16 per cent oxygen and the carbon dioxide is increased over a hundredfold to 4 per cent. In addition expired air is saturated with water vapour – this invisible loss of water from the body is about 1 litre a day.

## Nervous control of respiration

The respiratory centre, in the medulla oblongata of the brain, automatically controls the breathing muscles through the phrenic nerve to the diaphragm and through the many intercostal nerves to the intercostal muscles. It can also make use of the accessory muscles of respiration when they are required.

The respiratory centre has two sources of information about breathing: The degree to which the lungs are stretched and the level of carbon dioxide in the blood.

The stretching of the lungs through inspiration is detected by branches of the vagus nerve which reflexly inhibit inspiration and allow expiration to take place. The fluctuations in the level of carbon dioxide are detected by nerve endings in the aorta and carotid arteries. An increase in carbon dioxide causes an increase in acidity of the blood and this stimulates the respiratory rate – hyperpnea. Voluntary, rapid over-breathing decreases the carbon dioxide level in the body and this is followed by a period of cessation of breathing known as apnea. Permanently raised carbon dioxide levels, which can occur in some forms of lung and heart disease, lead the respiratory centre in the brain to adjust the respiratory rate to normal.

## Breathing and swallowing

Normally the vocal cords are open as breathing takes place. When something is swallowed the soft palate is pulled up and closes the post-nasal cavity. At the same time the epiglottis closes the top of the glottis by muscles pulling up the larynx.

## Involuntary breathing – cough

Irritation of the bronchi causes a deep inspiration followed by closing of the glottis. The expiratory muscles then contract so increasing pressure within the lungs. The glottis opens and an explosion of air, the cough, occurs.

## Sneeze

Irritation of the nasal mucosa produces a reflex like that of the cough, but, as the air is violently expelled from the lungs, the tongue blocks the back of the mouth and the air passes through the nose.

## Hiccup

A hiccup is due to a spasmodic contraction of the diaphragm causing the air to be rapidly inhaled. As this happens the glottis suddenly closes giving the characteristic noise.

## Emotional breathing – laughing and crying

Both are long inhalations followed by short, sharp exhalations.

## The organs of speech

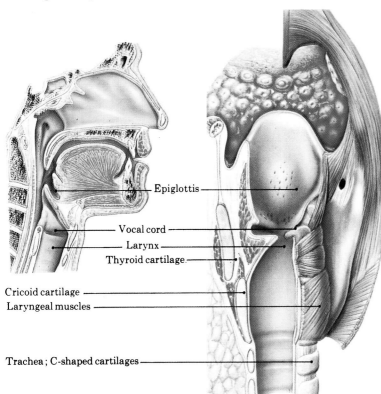

Epiglottis

Vocal cord

Larynx

Thyroid cartilage

Cricoid cartilage

Laryngeal muscles

Trachea; C-shaped cartilages

## Speech and voice mechanism

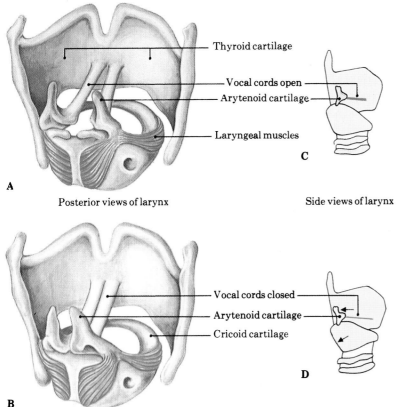

Thyroid cartilage

Vocal cords open

Arytenoid cartilage

Laryngeal muscles

**A**

**C**

Posterior views of larynx

Side views of larynx

Vocal cords closed

Arytenoid cartilage

Cricoid cartilage

**B**

**D**

### Yawning and sighing
Both are slow, long inhalations and gradual exhalation.

### Speech and voice production
The larynx (or voice box) is the complex cartilagenous structure between the pharynx and trachea. It divides the respiratory passage from that used for swallowing. It has two functions – to prevent food entering the trachea by closing the glottis; and to produce sound.

The two vocal cords are membranes that run backwards inside the larynx (*left*). Normally they remain open and still during breathing (**A**). When they are drawn towards each other by the laryngeal muscles, the passage of air will make them vibrate and produce sounds (**B**). The faster the air passes them the louder the sound. The tighter the cords are drawn together the higher the note (**D**), the slacker, the lower (**C**). The tension in the vocal cords is altered by the laryngeal muscles tilting the small, posterior laryngeal cartilages and arytenoid cartilage.

Voice production is a complex matter of co-ordinating the breathing muscles, the vocal cords and the lips and tongue. The vocal cords can produce a wide range of musical notes. The resonance of the notes depends on the shape of the chest, mouth and sinuses. The finer qualities of speech depend on the shaping of the tongue and lips to give the characteristics of sibilants, consonants and labials.

The voices of children of both sexes have a similar range of notes. At puberty, boys' voices "break". The larynx enlarges, due to the effect of the male hormone testosterone, and the vocal cords lengthen. This means that the boy is unable to produce high notes but now has a lower bass range. While the larynx is changing shape the quality of the voice also changes.

Speech depends on the brain. The production and quality of sound depends on the larynx and the surrounding structures to give resonance. The laryngeal nerves connect the "speech centre", in the cortex with the larynx. The speech centre is under the control of the cerebral hemispheres.

■ **Laryngitis**
Hoarseness of the voice due to inflammation of the vocal cords and the larynx.

■ **Pharyngitis**
Inflammation of the pharynx – back of the throat.

■ **Pleurisy**
Inflammation of the pleura.

■ **Pneumoconiosis**
Damage to the lung tissue from inhalation of dust particles, e.g. coal or stone dust.

■ **Pneumonia**
Infection of the alveoli causing impairment of breathing. This may either have spread from the bronchi, bronchopneumonia, or occur in the whole of the lobe, lobar pneumonia.

■ **Pneumothorax**
Air in the pleural cavity that may occur after the rupture of a lung.

■ **Rhinitis**
Inflammation of the nasal passages. This may be due to a coryza or to allergy, hay fever.

■ **Sinusitis**
Inflammation of the mucous lining of the sinuses.

■ **Tonsillitis**
Infection of the tonsils.

■ **Tracheitis**
Infection of the trachea.

■ **Tuberculosis** An infection of the lung caused by the organism of tuberculosis. This is a chronic illness that used to be a common occurrence.

**Cross-references**

Skeleton 20–3
Muscles 26–9
Heart 30–3
Blood 34–7
Nervous System 56–63

# Digestion – the Stomach

## The digestive tract – the mouth

Food has to be broken down into its basic molecules before it can be absorbed into the body.

The mouth (*right*) has four functions: It breaks up food by chewing with the teeth and tongue; it lubricates it with saliva, to make swallowing easy; it regulates temperature by either cooling or warming the food; it consciously initiates swallowing when the bolus is ready.

There are three pairs of salivary glands: Sub-mandibular; sub-lingual; and parotid. Between them they produce 1,500 to 2,000 millilitres of saliva a day, containing the starch-reducing enzyme ptyalin.

An adult has 32 teeth: of which there are 8 incisors; 4 canines; 8 premolars; 8 molars; and 4 wisdom teeth.

The tongue is an immensely mobile mass of muscle which helps the teeth to tear hard food into pieces by forcing it against the bony palate.

The throat and the esophagus connect the mouth with the stomach.

**The mouth**

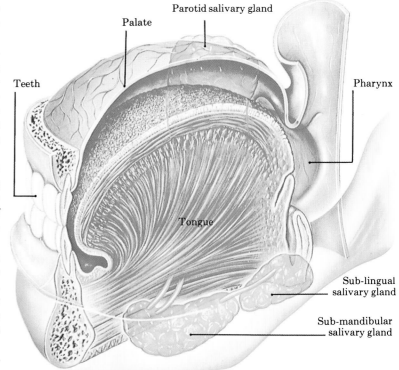

Palate
Parotid salivary gland
Teeth
Pharynx
Tongue
Sub-lingual salivary gland
Sub-mandibular salivary gland

**The pancreas**
magnified about 4 times

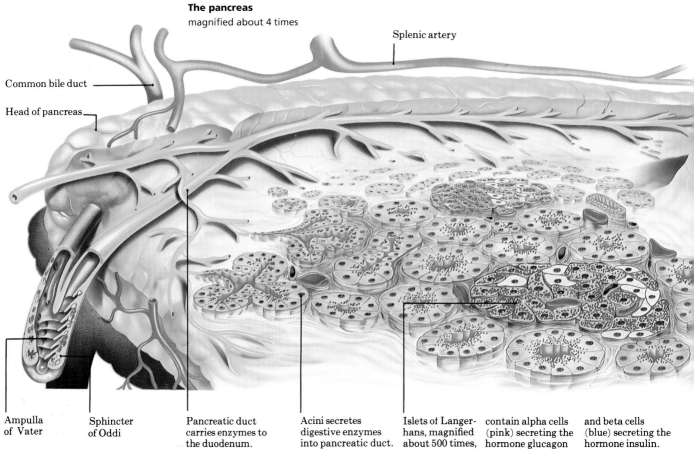

Splenic artery

Common bile duct

Head of pancreas

Ampulla of Vater

Sphincter of Oddi

Pancreatic duct carries enzymes to the duodenum.

Acini secretes digestive enzymes into pancreatic duct.

Islets of Langerhans, magnified about 500 times,

contain alpha cells (pink) secreting the hormone glucagon

and beta cells (blue) secreting the hormone insulin.

**The digestive tract**

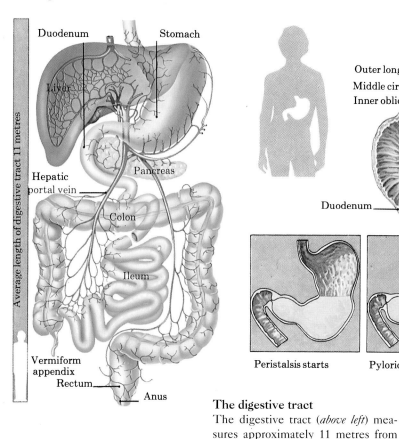

Duodenum  Stomach

Liver

Hepatic portal vein

Colon

Ileum

Pancreas

Vermiform appendix

Rectum

Anus

*Average length of digestive tract 11 metres*

Tail of pancreas

Mesenteric veins, containing insulin and glucagon.

**The stomach**

Outer longitudinal muscle
Middle circular muscle
Inner oblique muscle
Pylorus
Peritoneum
Duodenum
Mucosal lining

Peristalsis starts    Pyloric sphincter opens    Chyme enters duodenum

### The digestive tract

The digestive tract (*above left*) measures approximately 11 metres from mouth to anus. The first part, the oesophagus, is a strong, muscular tube in the posterior part of the thorax. It carries the bolus of food, by peristalsis, through the lower sphincter into the stomach.

### The stomach

The stomach (*above right*) is a muscular sac that can contain about 1.5 litres of fluid. It lies in the upper part of the abdomen under the liver and left part of the diaphragm. It has three activities: The storage and gradual release of food into the duodenum; the physical activity that churns and squeezes the semi-liquid food, chyme; and digestion.

The stomach has six kinds of secretion: Mucus, to act as a protective layer; hydrochloric acid, to sterilize the contents, neutralize the salivary enzyme – ptyalin – and break down the inactive enzyme – pepsinogen – into pepsin; pepsin, which splits protein into peptones; rennin, which curdles milk; intrinsic factor, for absorption of vitamin $B_{12}$ in the small intestine; and gastrin, to maintain gastric secretions.

Although the stomach is essentially a storage organ it is able to absorb some water, alcohol and glucose. Small amounts of chyme are passed on at a time into the duodenum, where the main part of enzyme digestion takes place. These enzymes are formed by the intestinal

### The pancreas

In the pancreas (*left*) the head and body lie in the loop of the duodenum, and the tail stretches across to the spleen. It consists of a number of lobules with ducts merging into a main pancreatic duct that joins the common bile duct. It then enters the second part of the duodenum at the ampulla of Vater.

The pancreas has two kinds of secretion: Insulin, which is secreted into the circulation and controls carbohydrate metabolism; and enzymes that enter the duodenum.

There are three pancreatic enzymes: Amylase, to turn starch into maltose; lipase, to split fats into fatty acids and glycerol; trypsin, to split peptones and proteins into amino-acids.

■ **Bilirubin** Principal bile pigment.

■ **Canaliculi** Channels in the liver lobules into which bile is secreted.

■ **Cholecystokinin (Pancreozymin)** Hormone produced by the intestinal wall to stimulate the contraction of the gall bladder and secretion of pancreatic enzymes.·

■ **Cystic duct** Tube from the gall bladder to join the hepatic and bile ducts.

■ **Enterogastrone** Hormone secreted by the duodenum to slow gastric motility.

■ **Enterokinase** Secreted by the duodenum to change inactive pancreatic trypsinogen into trypsin.

■ **Enterokinin** Duodenal hormone that maintains the duodenal secretions.

■ **Erepsin** Duodenal enzyme to break peptones into amino-acids.

■ **Fibrinogen** Blood protein concerned with clotting.

■ **Gastrin** Hormone produced by the stomach to maintain the gastric secretions.

■ **Glucose** Simple sugar.

■ **Glycogen** Simple form of starch that can be broken down into glucose.

■ **Hepatic duct** Tube draining the bile from the liver to join the cystic and bile ducts.

■ **Ileocaecal valve** Valve that allows the chyme to pass from the terminal ileum into the caecum but not to return.

■ **Insulin** Glucose-controlling hormone secreted by the pancreas.

■ **Intrinsic factor** Substance secreted in the stomach that combines with vitamin $B_{12}$, extrinsic factor, so that it can be absorbed by the villi in the small intestine.

■ **Invertase** Duodenal enzyme to reduce sugar to glucose.

# Digestion – the Intestines

## The digestive hormones and autonomic nervous system

The digestive secretions depend on a complex relationship between the autonomic nervous system and the local production of hormones. The sight or smell of good food will produce reflex peristalsis, salivation and gastric secretion that are then maintained when it is eaten.

Similar hormones to gastrin, from the gastric mucosa, are produced in the duodenum. Secretin stimulates the pancreas to produce an alkaline secretion to neutralize the acid chyme from the stomach. Fats cause secretion of cholecystokinin – pancreozymin, which makes the gall bladder contract and the pancreas secrete enzymes. Enterogastrone slows down gastric motility and speed of emptying. The presence of food also produces enterokinin to maintain the duodenal secretions.

## The small intestine

The small intestine (*right*) stretches from the pyloric sphincter to the ileocecal valve. It is about 7 metres long and is lined with small finger-like protuberances – villi – which are covered with even smaller ones – microvilli. This gives a total surface area of about 350 square metres for absorption. This areas sheds about 125 grammes of dead cells a day. Most of this loss of protein is reabsorbed into the body, but the debris is excreted. These villi are able to cope with 12 litres of intestinal contents, food, water and secretions, a day. The successful absorption of nutrients depends on their transport away from the cells of the villi into the bloodstream or lacteals.

**Small intestine**
magnified about 5 times

Duodenum is the short C-shaped section joining the stomach and jejunum.

Muscle layers have an internal circular and external longitudinal layer with autonomic nerves between them.

Hepatic portal vein drains digested food to the liver.

Branch of a mesenteric artery

Enzyme-producing glands

## The villus

Lacteal

Amino-acids
Glucose

Blood vessels

Fats

## The villi

Each villus (*left*) contains a network of blood vessels and a lymphatic vessel, the lacteal. The venous blood in the portal system takes the amino-acids, glucose, salts and water-soluble vitamins to the liver. The fatty acids and glycerol are combined together in the cells of the villus to form fats. These fats and the fat-soluble vitamins A and D drain into the lymphatic system via the lacteals in each villus.

Jejunum and ileum are two subdivisions of the small intestine.

Common bile duct

Pancreatic duct
Ampulla of Vater

Villi are finger-like projections lining the small intestine.

## The duodenum

The duodenum is the first C-shaped part of the small intestine and is responsible for the breakdown of food. The jejunum and ileum, extending from the duodenum, deal with absorption of food.

The duodenum produces three enzymes: Erepsin breaks peptones into amino-acids; invertase and maltase, to reduce sugar to glucose; enterokinase, to convert the inactive trypsinogen, the pancreatic secretion, to the active trypsin.

The bile salts, from the liver, help emulsify fats so that lipase can work more efficiently.

## Intestinal musculature

The large and small intestine have a mucosal lining, inner circular and an outer longitudinal muscle coat. In the large intestine the outer layer is present only as three bands – the teniae coli. The mucosa of the small intestine has patches of lymphoid tissue – Peyer's patches – which help eradicate infection.

## The large intestine

The large intestine is about 1.5 metres long. It starts at the cecum in the iliac fossa, ascends to the liver, then crosses the abdomen, as the transverse colon, to the spleen before descending to the pelvis. It then narrows to become the sigmoid colon and finally the rectum before ending at the anus.

The vermiform appendix is attached to the cecum. It has no known use. The liquid ileal contents pass through the ileocecal valve into the cecum. As these waste products pass down the colon water and salts are reabsorbed. The fecal matter is further decomposed by bacterial action to produce gas, some of the vitamin B complex and feces.

Normally the rectum remains empty, but the feeling of distension on filling with feces causes defecation. This is a reflex action in babies that comes under conscious control as the nervous system matures. Rectal filling occurs due to peristalsis and the more fibre in the diet the softer the feces and the easier it is to eventually defecate.

■ **Sinusoid** Channel taking the mixed blood from the hepatic artery and portal veins over the hepatic cells to the central vein of a liver lobule.

■ **Sphincter of Oddi** Circular muscle adjacent to the ampulla of Vater closing the joint pancreatic and biliary ducts

■ **Trypsin** Pancreatic enzyme that splits peptones and proteins into amino acids.

■ **Trypsinogen** Inactive pancreatic enzyme that is changed into trypsin.

■ **Urea** Nitrogen-containing substance formed from ammonia by the liver.

■ **Villus** Finger-like protuberance that increase the surface area of the small intestine.

■ **Vitamin B$_{12}$** An essential vitamin for the formation of red blood cells, the extrinsic factor, which combines with the intrinsic factor to be absorbed in the small intestine.

**Common diseases**

■ **Appendicitis** Inflammation of the vermiform appendix that frequently leads to perforation and peritonitis.

■ **Cancer** May occur anywhere in the digestive system. It is rare in the small intestine but common in the mouth and oesophagus, stomach and large intestine.

■ **Cholecystitis** Inflammation of the gall bladder.

■ **Cholelithiasis** Formation of stones in the biliary tract, principally the gall bladder. These may be mainly of bile pigment or cholesterol or a mixture of both.

■ **Cirrhosis** Scarring of the liver that ultimately leads to a raised pressure in the portal venous system, distension of blood vessels at the end of the oesophagus and causes haemorrhoids and ascites.

# Digestion – the Liver

## The liver

Lying mainly in the right side of the upper abdomen under the diaphragm and ribs, the liver (*right*) weighs about 1.5 kilogrammes. It is surrounded by a strong fibrous capsule that on the lower surface is covered with peritoneum. It has two lobes, a larger right and a smaller left. The gall bladder lies under the right lobe.

The liver contains 50 to 100,000 lobules. Each lobule has a central vein that drains blood into the hepatic veins. The portal veins and hepatic arteries are situated around the edge of the lobule and their blood intermingles to pass into the sinusoids that bathe the hepatic cells.

In the lobules, bile canaliculi drain bile into the branches of the bile duct, and thus into the hepatic duct. The bile is stored in the gall bladder, which contracts when stimulated by cholecystokinin. It then passes down the cystic duct into the common bile duct before joining the pancreatic duct at the ampulla of Vater.

The liver cells process the digested food, change it into substances that the body will need, then store them until they are required. If these substances are needed they will pass out of the cell into the sinusoid and thus to the central vein. Fats are either stored in the liver or sent back into the circulation of the fat depots. More frequently they are combined with protein – lipoprotein – to be used as an energy source. Proteins – the albumen, globulin and fibrinogen for the plasma – are all made in the liver. Glucose is turned into glycogen and iron combined with protein before being stored in the liver. The vitamins A, B complex, $B_{12}$, D, E and K are all also stored, the reserves lasting many months.

Liver cells also reprocess the body substances, such as haemoglobin, extracting the iron and re-using the amino-acids. Some of the amino-acids are changed into carbohydrates, to give readily available energy. The highly toxic substance ammonia is produced, which is turned into urea before being excreted by the kidneys. The heat produced by all this metabolism helps to keep the body temperature constant.

Bile canaliculus collects bile from hepatic cells.

## Microanatomy of the liver
magnified about 500 times

Hepatic lobule is the structural unit of the liver.

Central vein joins the hepatic vein.

Sinusoid is a vein through which blood flows over liver cells from hepatic to central vein.

## The liver's blood supply

Hepatic vein
Bile duct
Cystic duct
Hepatic artery

Stomach

Spleen

Portal vein

Branch of bile duct
Liver
Gall bladder
Pancreas
Duodenum
Common bile duct

Portal venule brings absorbed food from intestine.

Hepatic arteriole carries oxygenated blood.

Hepatic cell stores and processes absorbed food.

## Bile salts

Bile is a solution of cholesterol, bile salts and pigments. Bile emulsifies fats in the duodenum and is essential for the action of lipase and the absorption of fat-soluble vitamins. About 500 millilitres are produced a day. Some of it is reabsorbed, while the rest gives the particular colour to the feces.

## The blood supply

The liver has two blood supplies (*above*), 20 per cent comes from the normal systemic circulation bringing oxygen, while 80 per cent is the portal venous system draining the intestinal tract from the stomach to the rectum. This brings absorbed food to the liver cells. Over a litre of blood a minute passes through the liver.

■ **Mumps** Virus infection of the salivary glands that may also involve the pancreas.

■ **Pancreatitis** Inflammation of the pancreas often due to excess alcohol or to blockage of the ampulla of Vater by a gallstone.

■ **Peptic ulcer** Ulceration of the stomach or first part of the duodenum from excessive acidity or inadequate resistance of the mucosal lining.

■ **Peritonitis** Infection of the peritoneal cavity.

■ **Piles** *See* Haemorrhoids

■ **Ulcerative colitis** Inflammatory disease within the colon for which no infection has been found. It is a chronic and often severe disease.

■ **Volvulus** Twisting of the intestine on itself. This is most commonly seen in the sigmoid or pelvic colon.

**Cross references**
Muscles 29
Heart 30
Endocrine 53
Taste 73
Growth 86
Nutrition 90–3

# Urinary System – the Kidneys

**The urinary system – the kidneys**

The body has five organs of excretion: The skin for water, salts and urea; the lungs for carbon dioxide and water; the liver for bile salts, pigment and bilirubin; the intestine for roughage, water, salts and dead cells; and the kidneys. Each kidney is approximately 12 centimetres long, 3 centimetres thick and 7 centimetres wide, and weighs about 135 grammes, and is the most important route of excretion. A constantly adjusted trickle of urine is produced from the blood that enters each kidney, which is released when the bladder becomes full.

The kidneys filter 20 per cent of the blood supply at every heart beat. Blood under high pressure is forced into the capillaries of approximately 1 million glomeruli – Bowman's capsules – the goblet-shaped collecting cups of each filter unit – the nephron. The pressure forces 20 per cent of the plasma into the nephron, in the form of water, salts, urea, glucose and smaller molecules. Larger protein molecules and blood corpuscles cannot be forced through. The nephron then runs in a great U-shaped turn into the medulla. It consists of the proximal tube – loop of Henle – and distal tubule. It is surrounded by the blood vessels descending from the glomerulus. Two processes occur in the tubule: first, the lining cells reabsorb all the glucose, many of the salts and most of the water present in the filtrate; second, waste salts are secreted in the urine.

Urine is a pale yellow, slightly acid solution of salts, urea, uric acid, creatinine and metabolized hormones. Dilute urine will be less than the plasma's specific gravity of 1010, while concentrated urine may reach 1030. By selective excretion of acid or alkaline salts the slight alkalinity of the blood is kept constant.

The kidney has some part to play in the maintenance of blood pressure by the release of a hormone, renin, which increases blood pressure if the renal blood flow is reduced. The antidiuretic hormone, produced by the anterior pituitary gland, helps to regulate the reabsorption of water from tubules back into circulation.

**The kidney** magnified about 6 times

The renal vein carries filtered blood away from the kidney.

The renal artery conducts blood into the kidney to be filtered.

The loop of Henle is the central region of the nephron. Here reabsorption of essential materials from the filtrate into the bloodstream occurs.

The medulla, the inner region of the kidney, contains the loop of Henle and the collecting ducts surrounded by blood capillaries.

A capsule of tough, fibrous tissue surrounds and protects the kidney.

The ureter is a muscular tube about 25 cm long, draining urine to the bladder.

The pelvis, at the core of the kidney, acts as a funnel, collecting urine from the nephrons and conducting it to the ureter.

An arteriole supplies the glomerulus in the kidney's cortex with blood under high pressure.

The glomerulus is a tight knot of blood capillaries from which water and chemicals filter into the nephrons.

The cortex is the outer region of each kidney. It contains the glomeruli of each nephron and the supplying blood vessels.

Bowman's capsule —the expanded beginning of the tubule—is tightly clasped around the glomerulus. It is the nephron's collecting cup for fluid filtering from the blood.

The nephron is one of over a million coiled tubules in each kidney. It extends through the medulla as the loop of Henle, ending in the collecting tubule. Each nephron collects filtered fluid from the Bowman's capsule.

Collecting duct from nephrons.

Capillaries surrounding loop of Henle.

# Urinary System – the Bladder

## Organs of excretion

The urinary system (*below*) consists of two kidneys, the renal arteries and veins, two ureters and the bladder with the urethra.

The ureters, of circular and longitudinal peristaltic involuntary muscle, are 25 centimetres long, running behind the peritoneum and entering the posterior wall of the bladder at the base of the trigone.

The bladder, which can contain up to 500 millilitres of urine, lies in the pelvis, in front of the rectum and behind the symphysis pubis. In women it lies in front of the uterus.

Partly covered with peritoneum, it is supported by the muscles and ligaments of the pelvic floor. The muscle is formed from strong, interwoven, longitudinal transitional and circular fibres and lined with epithelium; it is controlled by the sacral nerves and autonomic nervous system.

The urethra is a fibro-elastic tube. In men it is about 20 centimetres long, passing through the prostate gland, pelvic floor and the corpus spongiosum of the penis. In women it is about 4 centimetres long, adjacent to the anterior wall of the vagina opening, behind the clitoris.

**Organs of excretion**

Kidney

One of two Renal veins

renal arteries

Ureter

Bladder

Prostate

Sphincters

Urethra

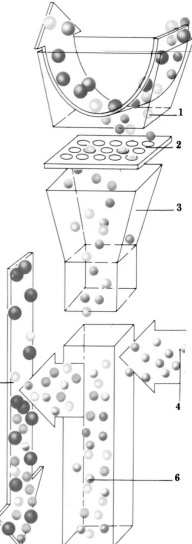

## The filtering mechanism

Blood flows at high pressure (*above*) through the capillaries of the Bowman's capsule (**1**) and only small molecules are forced through the walls (**2**) into the first part of the nephron (**3**). The filtrate passes down the proximal tubule, which secretes further metabolites and salts (**4**) and reabsorbs water, sodium, essential salts, glucose and amino-acids into the blood (**5**). The loop of Henle and distal tubule are concerned with the reabsorption of water and maintenance of the overall acid-alkaline balance of the body. Unwanted salts, urea and water are left as urine (**6**).

**The endocrine system**

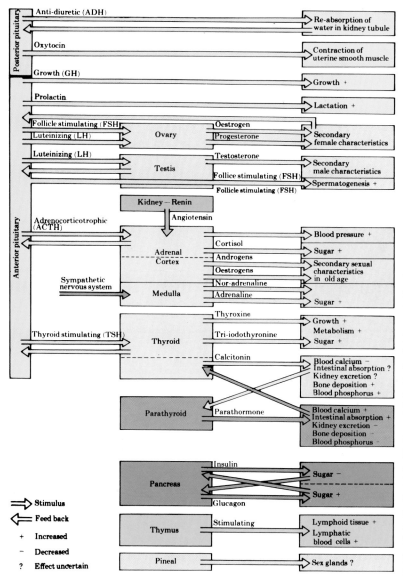

| Stimulus |
| Feed back |
| + Increased |
| − Decreased |
| ? Effect uncertain |

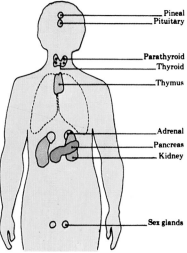

## The endocrine system

Hormone responses, transmitted through the circulation, are much slower than those produced by the nerves. Each hormone stimulates or suppresses certain cells because it is able to fit into the cell's metabolism like a piece in a jig-saw puzzle. The response is detected by the hypothalamus in a "bio-feedback" system and this modifies the production activity of the "master gland" – the pituitary gland. As can be seen in the diagram (*above*) this gland controls many, but not all, of the remaining glands. Glands often interact through effects on the metabolism.

## The lesser endocrine glands

The chart (*left*) shows the relationships between the various endocrine glands. Some of these have few or very local effects, such as the intestinal glands, while the activity of others, such as the pineal gland, is still not fully understood.

In the kidney renin is secreted in response to lowered blood pressure to combine with blood to form angiotensin, which stimulates aldosterone production.

In the intestine there is a variety of hormones with local effects: Cholecystokinin, enterogastrone, enterokinin, gastrin, pancreozymin and secretin.

The pineal body is a small structure adjacent to the roof of the third ventricle of the brain which may be associated with the development of testes and ovaries.

The thymus lies behind the upper part of the sternum and, until adolescence, plays an important part in the development of immunity by secreting a hormone to stimulate the production of lymphoid tissue and lymphocytes.

The pancreas produces exocrine digestive enzymes and two endocrine hormones, insulin and glucagon, from over one million islets of Langerhans. Insulin is produced by the beta cells of the islets of Langerhans in response to rising blood sugar levels and acts by increasing storage and use by the tissues while reducing liver production. It also reduces the use of fat by the body. Glucagon, from the alpha cells of the islets of Langerhans, responds to a low blood sugar by increasing glucose production. Glucose levels are also affected by both the thyroid and growth hormones – adrenaline, nor-adrenaline and cortisol.

The testes respond to follicle stimulating hormone by the production of sperm and luteinizing hormone by increased testosterone levels.

In the ovaries follicle stimulating hormone causes maturation of the ovarian follicle with production of oestrogen. Luteinizing hormone produces the post-ovulatory secretion of progesterone. The balance is achieved by "bio-feedback".

### Definitions

■ **Adrenaline** Hormone that is produced by the adrenal medulla. It increases the cardiac rate and blood glucose level.

■ **Aldosterone** Mineral-corticosteroid produced by the adrenal cortex.

■ **Angiotensin** Peptide produced by the production of renin and blood to stimulate aldosterone production and increase sodium reabsorption.

■ **Antidiuretic hormone (ADH) – vasopressin** Hormone produced by the posterior lobe of the pituitary gland that acts on the kidneys to increase the reabsorption of water.

■ **Calcitonin** Calcium-controlling hormone produced by the thyroid gland.

■ **Cortisol** Glucocortico-steroid hormone produced by the adrenal cortex.

■ **Endocrine** Gland that secretes a hormone directly into the blood stream.

■ **Exocrine** Gland that secretes enzymes into a duct.

■ **Glucagon** Glucose-stimulating hormone produced by the alpha cells of the Islets of Langerhans of the pancreas.

■ **Gluco-cortico-steroid** One of a group of adrenal hormones, principally cortisol, which stimulates carbohydrate metabolism.

■ **Growth hormone** Hormone produced by the anterior lobe of the pituitary gland.

■ **Hydrocortisone** See Cortisol.

■ **Insulin** Hormone produced by the beta cells of the Islets of Langerhans that reduces the glucose level in the blood.

■ **Mineral-cortico steroid** Adrenal hormones, mainly aldosterone, controlling salt reabsorption in the kidneys.

# Endocrine Glands

**The pituitary gland**

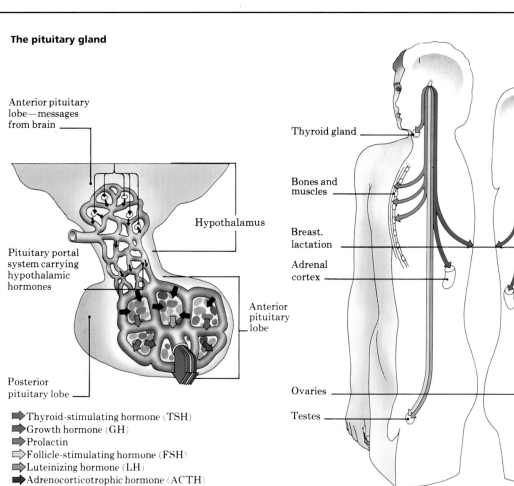

Anterior pituitary lobe—messages from brain

Hypothalamus

Pituitary portal system carrying hypothalamic hormones

Anterior pituitary lobe

Posterior pituitary lobe

Thyroid-stimulating hormone (TSH)
Growth hormone (GH)
Prolactin
Follicle-stimulating hormone (FSH)
Luteinizing hormone (LH)
Adrenocorticotrophic hormone (ACTH)

Thyroid gland
Bones and muscles
Breast. lactation
Adrenal cortex
Ovaries
Testes

## The pituitary gland

The pituitary gland (*above*) lies under the centre of the brain, protected by the pituitary fossa – sella turcica – of the sphenoid bone. It is composed of two lobes, which measure 1 centimetre across and are connected to the base of the brain, just behind the optic chiasma, by a short stalk. The anterior lobe, about 70 per cent of the total gland, originally derives from the endoderm layer of the embryo and the posterior lobe from the ectoderm. This is reflected in their different functions.

The circadian biological rhythms – the emotional and physical changes – as well as the cerebral hemispheres, may stimulate or suppress the activity of the hypothalamus. The hypothalamus secretes controlling hormones into the pituitary portal venous system to increase or decrease hormonal activity in the anterior lobe.

The anterior lobe produces two kinds of hormones – those hormones that cause a direct response in the body and the trophic hormones that control the activities of the other endocrine glands. The level of stimulus from the trophic hormones is directly related to the response of the "target" gland and its "feedback" to the hypothalamus that assesses the response. This achieves a balanced hormonal activity.

Of those hormones that affect the body directly, growth hormone (GH) stimulates the production of somatomedin by the liver which stimulates growth of bone, cartilage and possibly other body tissues. Prolactin helps to produce lactation and suppresses ovulation in the lactating woman.

Thyroid stimulating hormone (TSH) increases thyroxine and tri-iodothyronine production from the thyroid gland.

Follicular stimulating hormone (FSH) and luteinizing hormone (LH) control the production of spermatozoa and testosterone in the testes, and oestrogen and progesterone in the ovaries.

Adreno-corticotrophic hormone (ACTH) stimulates gluco- and mineralo-corticosteroid production from the adrenal cortex.

The posterior lobe is like a physical extension of the hypothalamus as it is under nervous control. It secretes two hormones, which are stored until they are needed.

The first, antidiuretic hormone (ADH), also known as vasopressin, acts on the tubule of the kidney nephron to increase the reabsorption of water. Any change in the osmotic pressure of the blood is assessed by the osmoreceptors in the hypothalamus; an increase causes stimulus to the thirst centre and release of ADH. There is an automatic increase in ADH secretion at night. The other hormone, oxytocin, stimulates the onset of labour, affects smooth muscle and helps in lactation.

## Adrenal glands

The two adrenal glands (*right*) lie on the top of the kidneys and have a large blood supply as well as sympathetic nerves. There are two layers, the cortex and the medulla, through which a rich arterial supply reaches the underlying veins. Lying on the medulla of irregular cells surrounding the venous sinuses are long, straight columns of cells – zona fasciculata – followed by a layer of spherical cells – zona glomerulosa.

The adrenal cortex produces three kinds of hormones: Gluco-corticosteroids, mainly cortisol (hydrocortisone); mineralo-corticosteroids, mainly aldosterone; and a mixture of androgens and estrogens.

Cortisol helps stimulate glucose, amino-acid and fat metabolism to repair the body and increase levels of blood sugar.

Aldosterone monitors the level of body salts and causes reabsorption of sodium and loss of potassium from the kidney tubule. Aldosterone also increases intestinal sodium absorption and decreases its loss in sweat. Angiotensin stimulates aldosterone.

Androgens and estrogens are produced in small amounts and continue

**The adrenal gland**

Adrenal arteries and veins

Adrenal gland

Kidney

to do so after the testes and ovaries cease developing. This feminizes men to a mild degree and masculinizes women when they are older.

ACTH stimulates the production of gluco-corticosteroids, and, to a far lesser extent mineralo-corticosteroids and sex hormones. It is controlled by the "bio-feedback" – the biological mechanism whereby the increase in production of a hormone causes a decrease in the production of the trophic hormone and thus reduces its own manufacture.

In stressful situations, such as emotional disturbance, pain, exposure to extremes of temperature or low levels of sugar in the blood, the adrenal medulla produces adrenaline and noradrenaline as a direct response to the sympathetic nervous system. The body is prepared for immediate action, commonly known as "flight or fight", with increased blood pressure, shift of blood from the intestines to muscles, rapid heart rate, dilitation of bronchioles, sweating and increased alertness.

**The parathyroid gland**

A

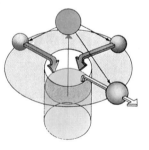

B

C

## Thyroid gland

The two lobes of the thyroid gland (**A**), joined by the isthmus, lie adjacent to the upper part of the trachea. The gland produces thyroxine and tri-iodothyronine, that respond to TSH. They stimulate cell metabolism, increase blood sugar and are necessary for normal growth. It also secretes calcitonin which lowers blood calcium levels by stimulating deposition in the bones. It counteracts the action of parathormone and may affect absorption of calcium from the intestine and excretion in the urine.

## Parathyroid glands

Four small glands are found in the substance of the thyroid gland (the black dots in **A**) and produce parathormone (PTH). This hormone regulates, with calcitonin, the levels of calcium and, indirectly, phosphate in the blood. Seen in **B**, low blood calcium (orange) increases the output of PTH to mobilize calcium from the bones (blue), inhibit secretion in the kidneys (red) and increase absorption from the intestines (green); this increases the blood calcium (C) and in turn inhibits the release of PTH while stimulating calcitonin.

**Common diseases**

■ **Acromegaly and gigantism** Due to an excess of growth hormone causing excess growth before maturity and thickening of bones and skull as an adult.

■ **Addison's disease** Deficiency of the adrenal cortical hormones leading to weakness, low blood pressure, vomiting and eventual death.

■ **Cranio-pharyngioma** Cystic tumour of the pituitary gland.

■ **Cretin** Deficiency of thyroid hormones at birth leading to stunted growth and imbecility.

■ **Cushing's disease** Excess production of cortisol causing raised blood pressure, and a characteristic moonlike appearance to the face with obesity.

■ **Diabetes insipidus** Frequent passing of large volumes of urine due to deficiency of the antidiuretic hormone.

■ **Goitre** Swelling of the thyroid gland. This may be due to deficiency of iodine or excessive production of thyroid hormone.

■ **Hyperthyroidism** Excessive secretion of the thyroid hormones causing loss of weight, anxiety, rapid heartbeat and sweating.

■ **Myxoedema** Deficiency of thyroid hormone leading to a slow mental state, thickening of the facial features, loss of hair and low body temperature.

■ **Phaeochromocytoma** Tumour of the medulla of the adrenal gland with excessive production of adrenaline and noradrenaline.

# The Nervous System

**Anatomy of the nervous system**

Cerebrum

Cranial nerves

Spinal cord

Cervical nerves (eight pairs)

Thoracic nerves (twelve pairs)

Lumbar nerves (five pairs)

Sacral nerves (five pairs)

Coccygeal nerves (one pair)

Sensory root

Spinal nerve

Vertebra

Motor root

Sympathetic ganglia of autonomic nervous system

## The anatomy of the nervous system

The nervous system keeps control of all the body's activities. It has two parts, the central and peripheral nervous systems (*above*). The central nervous system consists of the brain and the spinal cord. The peripheral carries information to the central nervous system through the afferent, sensory nerves and carries out instructions through the efferent, motor nerves. The autonomic nervous system works outside the main nervous systems.

## The brain – basic facts

The brain is a soft, jelly-like structure, weighing about 1,380 grammes in men and 1,250 grammes in women, surrounded by the meninges. It consists of ten thousand million neurons held in place by eighty thousand million neuroglia or glia.

There are three parts to the brain: The forebrain – the pair of cerebral hemispheres, thalamus, hypothalamus and limbic system; the midbrain – the first part of the brain stem; and the hindbrain – cerebellum, pons and medulla.

## The meninges

The brain is enclosed by three membranes – the meninges – within the skull. The tough, outer dura mater and the thin, inner pia mater sandwich, the arachnoid mater. The arachnoid mater is pierced by many channels and allows the circulation of the cerebro-spinal fluid around the surface and to the ventricles and central canal of the spinal cord. The meninges, which also carry blood vessels, surround the nerves as they leave the brain and extend over the spinal cord down the neural canal.

## Cranial nerves

There are twelve nerves, I-XII (*right*), which arise from the under-surface of the brain. They supply the head, neck and major organs in the body. Three are solely sensory (blue): Olfactory I; optic II; and auditory VIII. Two are entirely motor (red): Accessory to neck muscles XI; and hypoglossal XII; to the tongue. The rest are mixed: Oculomotor III; trochlear IV, and abducent VI, supply the eye muscles; trigeminal V, involves chewing and facial sensations; VII facial, involves facial expression and taste in the anterior two-thirds of the tongue; glossopharyngeal IX, involves swallowing and the remainder of taste; vagus X, supplies chest and abdominal organs.

## Neurons

Neurons are the basic unit of the nervous system. Afferent neurons carry impulses from the sensory receptors, efferent neurons carry instructions to the muscles or glands, and interneurons link other neurons. Neurons (*below*) consist of a cell body with nucleus and small dendrites that pick up messages. An axon connects with the next cell, where it breaks up into many small axon terminals. There is a gap – the synapse – between the axon terminal and the next cell. Neurons combine to form nerves.

**Cranial nerves**

I II IV VI III VII XII XI V VIII IX X

## Nerve conduction

Normally the nerve cell is potentially excitable – polarized – with potassium ions inside and sodium ions outside the membrane. Once excited there is an exchange of ions – action potential – and a wave of electrical excitation will pass down the axon to the synapse to release noradrenaline or acetylcholine. The nerve is momentarily depolarized and while the ions are returning across the membrane it is in a "refractory" state. The synapse chemicals are rapidly destroyed by local enzymes; the synapse is then ready for action.

**Neurons**

Nucleus

Cell body

Dendrite

Nodes of Ranvier

Schwann cell nucleus produces myelin

Axon

Myelin sheath

Nerves made up of hundreds of axons

Endoneurium sheath surrounds one nerve fibre.

Perineurium sheath surrounds bundles of nerve fibres.

Epineurium sheath surrounds the whole nerve.

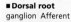

- **Dorsal root** ganglion Afferent nerves containing neuron nuclei.
- **Dura mater** Outer layer of the meninges.
- **Foramen of Magendie** Hole in the roof of the fourth ventricle connecting the subarachnoid space with the ventricular system.
- **Foramen magnum** Hole in the base of the skull.
- **Fornix** Part of the limbic system.
- **Glia** Supporting cells of the nervous system.
- **Grey matter** Areas of the central nervous system containing many neuron nuclei.
- **Gyrus** Fold of the cerebral cortex.
- **Hippocampus** Part of the limbic system.
- **Homunculus** Psychiatric term to describe a small man created in the imagination.
- **Hypothalamus** Concerned with nervous control of the pituitary gland and vital nerve centres.
- **Lentiform nucleus** One of the basal ganglia.
- **Limbic system** Pair of structures concerned with memory and the instinctive emotions.
- **Lobe** Anatomical division of the brain.
- **Mammillary body** Initial relay station of the limbic system.
- **Medulla oblongata** Part of the brain stem area connecting with the cerebellum.
- **Meninges** Three layers coating the nervous system.
- **Motor end plate** Specialized synapse between the axon terminals and muscle fibrils.
- **Myelin** Fatty coating of nerve.
- **Neural canal** Bony channel in which the spinal cord lies.
- **Neuroglia** *See* Glia
- **Neuron** Basic nerve cell.

# Nervous System – the Spinal Cord

■ **Node of Ranvier**
Minute constriction in the myelin sheath of a nerve.

■ **Nucleus** Collection of nerve cells with one function.

■ **Osmoreceptor** Receptor sensitive to osmotic change.

■ **Pia mater** Inner, and thinnest, layer of the meninges.

■ **Pons** Area of interconnecting nerve fibres below the cerebellum.

■ **Reflex arc** Nervous response involving central nervous system.

■ **Reticular formation** Series of short nerve fibres whose activity increases alertness.

■ **Schwann cell nucleus** Concerned with the formation of myelin for the neuron.

■ **Septum pellucidum** Part of the limbic system.

■ **Spinal cord** Extending from the foramen magnum to the second lumbar vertebra with 31 pairs of nerves.

■ **Sulcus** One of 200 grooves between gyri.

■ **Thalamus** Concerned with sensory nervous transmission.

■ **Ventricle** One of four brain cavities.

■ **Vital centre** Nerve nuclei concerned with the functioning of the body.

■ **White matter** Area of the brain containing interconnecting neurons between areas of grey matter.

**Common diseases**

■ **Cerebral haemorrhage** Bleeding from a cerebral artery into the surrounding brain tissue.

■ **Cerebral thrombosis** Blockage of a cerebral artery with blood clot.

■ **Cerebral tumour** Malignant or benign tumour may occur from glial cells or spread from other parts of the body.

**Spinal cord**

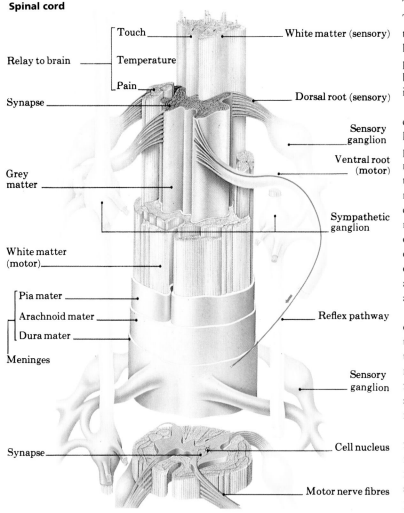

Touch
Temperature
Pain

Relay to brain

Synapse

Grey matter

White matter (motor)

Pia mater
Arachnoid mater
Dura mater

Meninges

Synapse

White matter (sensory)

Dorsal root (sensory)

Sensory ganglion

Ventral root (motor)

Sympathetic ganglion

Reflex pathway

Sensory ganglion

Cell nucleus

Motor nerve fibres

## Spinal nerves and reflex arc

The 31 pairs of spinal nerves (*right*) overlap each other in a way that developed in the embryo (*far right*). These nerves serve specific areas of skin known as dermatomes.

For the reflex arc to function five components are required: Sensory receptor; afferent nerve; interconnecting neuron; efferent nerve; and an end organ, such as muscle. Stimulation of the sensory nerve sends an impulse to the spinal cord; the impulse is relayed through at least one interconnecting neuron to activate the efferent nerve.

Consciousness is not initially involved in the reflex arc. This can be seen in the patellar reflex or in the emptying of a baby's full bladder. As the baby matures, conscious inhibition can prevent the bladder reflex but not the patellar reflex.

## The spinal cord

The spinal cord (*left*) extends from the foramen magnum to the second lumbar vertebra. It is covered with pia, arachnoid and dura mater and is bathed in cerebro-spinal fluid, which is also found in the central canal.

Sensory information reaches the cord through the dorsal nerve root known as the dorsal ganglion. The posterior half of the central "grey matter" relays the afferent impulses to the columns of fibres in the surrounding "white matter". These columns have long fibres, ultimately reaching the brain, and shorter fibres connecting nearby segments of the cord. The lateral columns are concerned with pain and temperature and the posterior columns with touch and proprioception.

Some neurons transmit faster than others. The sensory nerves contain thicker myelinated fibres, for the fast transmission of pain at about 100 metres a second; non-myelinated fibres transmit heat or cold at slower speeds. The motor area is in the anterior half of the cord, with the central "grey matter" collecting instructions from the anterior columns and relaying them through the ventral nerve roots to the end organs. The dorsal and ventral roots combine to form 31 pairs of spinal nerves.

The autonomic nervous system runs parallel to the spinal column and interconnects with the spinal nerves. This nervous system controls such things as the reflex actions of alimentary smooth muscle, splanchnic blood flow and heart rate.

**Skin sensation**

Cervical (eight pairs)
Thoracic (twelve pairs)
Lumbar (five pairs)
Sacral (five pairs)

Front        Back

**Six-week embryo showing primitive dermatomes**

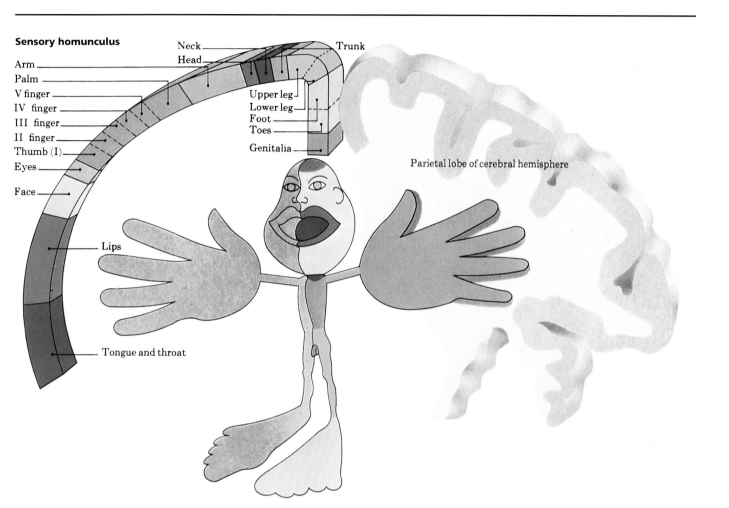

**Sensory homunculus**

Arm
Palm
V finger
IV finger
III finger
II finger
Thumb (I)
Eyes
Face
Lips
Tongue and throat

Neck
Head
Trunk
Upper leg
Lower leg
Foot
Toes
Genitalia

Parietal lobe of cerebral hemisphere

## Cerebral cortex – homunculus

This weird picture of a man (*above*) is a visual representation of the relative proportions of the neurons in the sensory area behind the main sulcus of the brain. Skin areas on the trunk and limbs are poorly supplied with sensory nerves, while the lips, hands and feet are extremely sensitive and are relatively huge.

The visual cortex is in the occipital lobe and the auditory area in the temporal lobe so they are not proportionately represented. A similar but differently shaped homunculus can be made for the motor cortex. The hands are still huge but the face and tongue, trunk and legs are larger with much smaller feet.

This representation shows the right hemisphere, which controls the body's left side. The coloured areas on the diagrammatic representation correspond in size to the coloured areas on the body.

## Intelligence

Intelligence is thought to be 80 per cent inherited and 20 per cent acquired. It can be approximately measured using the Stanford-Binet test in children up to the age of 15 and after that age the Wechsler test. Sex, race and cultural background have a variable bearing on the results. An average child who is well educated with a good understanding of words will appear more intelligent than is really the case. A rating of 100 is average and 130 very superior; below 70 is considered subbnormal.

There seems to be little relationship between intelligence and creativity. Intelligence tests tend to assess the left cerebral hemisphere as this is the logical half of the brain, while creativity and originality are more from the right cerebral hemisphere. These are more difficult to measure as they depend on the subjective interpretation of the psychologist.

## Learning

Learning is the way that new skills are acquired. Certain ways of doing something will produce a benefit or reward or, alternatively, something unpleasant so that repeating the action is avoided.

## Memory

Memory may be transient and immediately forgotten, such as not remembering everyone encountered in the street. Short-term memory is for things that are immediately necessary, such as shopping. Long-term memory can be "recalled" to consciousness when required.

## Consciousness

Consciousness is constant activity of the cerebral cells by the excitation of the recitular formation. The subconscious is the activity of the memory part of the brain which may, unknowingly, modify the conscious.

■ **Encephalitis**
Inflammation of the brain cells due to an infection.

■ **Epilepsy** Ill-defined episode of excessive activity in some part of the brain characterized by fits.
Grand Mal – the major convulsive fit
Petit Mal – a momentary loss of awareness without, however, loss of consciousness

■ **Hydrocephalus**
Condition in which there is increased pressure of the cerebrospinal fluid within the brain, causing distension and, in infants, ensuing enlargement of the skull and damage to the brain tissue.

■ **Meningitis**
Infection of the meninges by bacteria or viruses.

# Nervous System – the Brain

**The brain**

Optic chiasma

Mamillary body and septum pelucidum

Amygdaloid body

Parietal lobe

Frontal lobe

Occipital lobe

Temporal lobe

Cerebral cortex—grey matter

Cerebrum—white matter

Corpus callosum

Thalamus

Lymbic system

Hypothalamus

Pituitary gland

Medulla and the base of the fourth ventricle

Cerebellum

Lateral ventricles

Third ventricle

Fourth ventricle

Spinal cord

## The forebrain – the cerebrum

Seventy per cent of the brain is formed by the two cerebral hemispheres (*above*). They are linked together by the nerve fibres running in the corpus callosum. The surface area is increased thirty times by the sulci and gyri. The cortex is densely packed with nerve cells, and the white matter with nerve fibres.

The nerve fibres cross over in the brain stem so that the right hemisphere is concerned with the left side of the body and vice versa. The left hemisphere is primarily the speech centre and mainly controls logical behaviour, but has to rely on the right side to assess a three-dimensional world, appreciate artistic values and recognize friends. The frontal lobe has conscious control of movement. The prefrontal area is concerned with intelligence and personality. The parietal lobe concerns sensation and body position. The smaller occipital lobe assesses vision and the temporal lobe assesses hearing. In between are the association areas linking the cerebral cortex.

## The ventricles

Each cerebral hemisphere has a space known as a lateral ventricle. The two "arms" extend forwards into the anterior and temporal lobes and the "handle" into the occipital region. The lateral ventricles fuse between the two thalami to form the third ventricle. A central duct leaves the fourth ventricle to pass down through the spinal cord. The fourth ventricle lies between the cerebellum and brain stem. The ventricular system is filled with cerebro-spinal fluid.

## The cerebro-spinal fluid

Secreted by the choroid plexus, the cerebro-spinal fluid is clear and colourless. It bathes the surface of the brain in the subarachnoid space, between the arachnoid and pia mater, and internally in the ventricles, which it joins through a channel in the roof of the fourth ventricle – the foramen of Magendie.

It has three functions: To take food to the brain and spinal cord; to remove metabolites; to act as a shock absorber for the brain.

## The basal ganglia or nuclei

The most important nuclei are the caudate and lentiform in the corpus stratum which, with the thalamus and cerebellum, produce smooth, precise movement. They help to co-ordinate motor movements.

## The limbic system

Consisting of a pair of structures lying above the thalami, the limbic system links the midbrain with the hippocampus and cingulate gyrus and the cerebral cortex. It is concerned with instinctive emotions and memory. The mamillary body acts as the initial relay station through the fornix to the other areas. The septum pellucidum is concerned with pleasurable emotions, the amygdaloid body regulates anger and aggression. The hippocampus and cingulate gyrus help with memory by relating what is happening now with the memory of the past. This helps to maintain the brain's concentration without being distracted by useless information. This inhibits the "alertness" created by the reticular formation.

## The thalamus

Acting as a relay for sensory information, the thalamus sorts impulses and sends them to the appropriate areas of the parietal lobes. It is also believed to assess sensation.

## The hypothalamus

Lying below the thalamus and above the pituitary gland in the centre of the brain, the hypothalamus is partly under the control of the prefrontal lobes and connected with the limbic system. It acts as a link between the nervous and endocrine systems, regulating the autonomic systems as well as the metabolic state of the body.

It is connected with the anterior part of the pituitary gland by a portal venous system beginning and ending in capillaries; this is similar to the hepatic portal system. This blood carries the hormones released in the hypothalamus to control the release of the anterior pituitary hormones. The posterior pituitary gland is controlled by nerves.

The hypothalamus contains centres to control appetite, thirst, body temperature and libido. There is a fine balance between the amount of food eaten and used. Disturbance of the appetite centre may lead to excess appetite and weight gain.

If the blood becomes diluted the hypothalamus suppresses the production of the antidiuretic hormone (ADH) to allow diuresis to occur. Haemoconcentration will produce the opposite effect with stimulus to produce ADH from the posterior pituitary gland, and the sensation of thirst to stimulate the individual to drink. Libido is due to three factors: Instinctive responses, which are genetically inherited; the sex hormones; and sexual behaviour and responses learned from infancy. Hunger and thirst will reduce libido through its hypothalamic centre.

Temperature regulation is controlled through an assessment of the temperature of the blood. Heat loss can be increased by dilatation of the blood vessels in the skin, convection loss and sweating; it may be reduced by vaso-constriction and, if necessary, increasing heat production by shivering. The thyroid hormone, thyroxine, has a direct effect on the temperature control.

The activities of the hypothalamus integrate with the rest of the brain's activities, but are most closely correlated with those of the brain stem.

## The hindbrain – the cerebellum

The hindbrain is joined directly to the spinal cord. It contains the medulla – concerned with sleep, consciousness, breathing and blood circulation – and the cerebellum.

The cerebellum grows rapidly and almost reaches adult size by the age of two. It co-ordinates the movements initiated by the rest of the brain. This is done by information relayed to it from the medulla oblongata via the nerves in the pons. An assessment is made of the conscious desire; information on balance is received from the ears, visual assessment from the eyes, and other sensory information from the rest of the body. A smooth, regulated movement can then be made, mainly by the inhibition of overreactions.

■ **Stroke** Lay term that refers to the sudden interference with the circulation of blood in part of the brain causing the onset of paralysis usually accompanied by a temporary loss of consciousness. The paralysis may occur in the arm and leg on one side – hemiplegia; both legs – paraplegia; or all four limbs – tetraplegia.

■ **Subacute combine degeneration of the cord** Sensory loss and motor weakness due to vitamin B$_{12}$ deficiency.

■ **Subarachnoid haemorrhage** Term to describe bleeding from a burst blood vessel into the cerebrospinal fluid.

■ **Subdural haematoma** Blood clot, usually after injury, occurring beneath the dura mater.

## Areas and functions of the brain – memory, movement, emotion

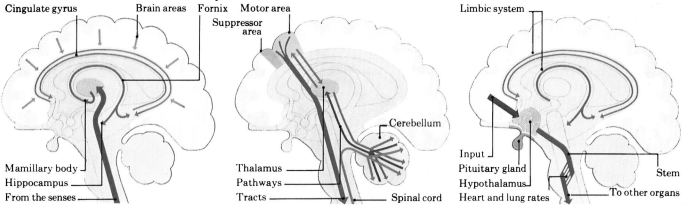

### Memory

Memory (*above*) begins with sensation from the body passing to the cerebral cortex. As the information is relayed through the thalamus it is also passed to the mamillary body and into the limbic system. The stimulus passes through the fornix to the hippocampus and then outwards into the diffuse area of the cingulate gyrus. If the memory of a similar stimulus is aroused, the cerebral cortex may be activated.

### Movement

Movement (*above*) is initiated in the motor cortex and immediately modified by the adjacent suppressor cortical area before direct transmission to the muscles. Each muscular movement is assessed and modified by the cerebellum in conjunction with the thalamus. An unconscious co-ordination of movement with position, balance and vision has to be made so that other muscles automatically adjust to give conscious movement.

### Emotion

The emotions (*above*) are a complex integration of conscious reaction, memory and instinctive desires. The frontal lobes and limbic systems both affect the hypothalamus, with its centres for anger, thirst, appetite and sexual desire. These may be stimulated to interreact with the motor activity in the brain stem causing alterations in heart, respiratory rate and muscle tone. The hypothalamus also stimulates the pituitary gland.

# Nervous System – the Brain Stem

## The midbrain – the brain stem

The brain stem (*right*) is the vital control area of the brain and is concerned with maintaining all the essential regulatory mechanisms of the body: Respiration; blood pressure; pulse rate; alertness and sleep.

In the medulla oblongata – the lower end of the brain stem – the nerve fibres from the spinal cord, both sensory and motor, cross over.

At the upper end there are links with the higher cerebral centres of consciousness – the limbic and thalamic areas. The nuclei that reflexly control pupil size and eye movements lie in this area of the midbrain.

The third and fourth ventricles with the interconnecting central canal bathe the upper surface with cerebro-spinal fluid formed by the choroid plexuses. The central canal is joined to the subarachnoid space by the foramen of Magendie.

The pons connects the brain stem with the balancing control of the cerebellum. Short nerve fibres – the reticular formation – link the vital centres with each other and monitor the information reaching the brain stem from the sensory tracts of the spinal cord. A change in body position will cause an alteration in blood pressure and pulse rate by control of the diameter of the arterioles. The activity of the reticular formation maintains wakefulness and alertness. If this activity is slowed, for example after a large meal in quiet surroundings, sleep may occur. Anxiety or fear will increase alertness and raise the blood pressure and pulse rate.

The reticular formation not only controls the vital centres, as well as the production of hydrochloric acid in the stomach, but also helps correlate information from the eyes and ears, and smell and taste from the tongue. The sight and smell of a good meal will cause salivation, increased gastric secretion and peristalsis. The reticular formation then controls the mechanism of swallowing.

The vomiting centre may be activated by nauseating tastes or, in some people, sickening sights. It is easily stimulated by movement – motion sickness – due to the constant changes in the organs of balance.

**Anatomy of the brain stem**
Magnified about 3 times

Reticular formation is a tangled mass of short-axoned neurons.

The spinal cord has central grey matter surrounded by white matter.

Sensory nerve bundles arise from the left side of the body and pass to the right side of the brain after crossing in the medulla.

Motor nerve bundles emerge from the right side of the brain and cross to supply the body's left side.

Motor and sensory fibres cross in the medulla to serve and supply areas on the opposite sides of the body.

The medulla oblongata is the lower part of the brain stem.

Motor nuclei of cranial nerves are shown on the left, coloured pink. This one is the oculomotor nucleus

The choroid plexus of the third ventricle secretes a clear, watery fluid —the cerebro-spinal fluid.

The thalamus receives both motor and sensory nerve fibres as they pass into and out of the brain.

The sensory nucleus of the fifth cranial nerve.

The brain stem is shown (*above*) in its actual position on the undersurface of the brain. The light brown oblong indicates the place through which the section (*left*) is taken.

The trigeminal nerve is comprised of single fibres that unite to form a nerve bundle.

The pituitary stalk, part of the forebrain, lies above the pons.

Motor and sensory nerve fibres pass through the brain stem on their way into and out of the spinal cord.

The pons is made up of transverse nerve fibres interwoven with longitudinal nerve fibres.

The optic nerve, which splits to enter the chiasma, lies in front of the brain stem.

# Nervous System – the Eye

## Definitions

■ **Anterior chamber** Front part of the eyeball bounded by the cornea in front and the lens and iris behind.

■ **Aqueous humor** Fluid contents of the anterior chamber.

■ **Bipolar cell** Cell with dendrites at both ends.

■ **Blind spot** Point at which the optic nerve enters the eyeball.

■ **Canthus** Angle where the eyelids join.

■ **Cerebellum** Part of the brain associated with co-ordination and balance.

■ **Chiasma** Area in which the two optic nerves cross.

■ **Choroid** Middle vascular coat of the eye.

■ **Ciliary body** Lies between the two chambers containing the ciliary muscle and ducts for draining the aqueous humor.

■ **Ciliary muscle** Muscle in the ciliary body to which the suspensory ligament is attached.

■ **Cones** Cells that respond to colour.

■ **Cornea** Transparent curved anterior part of the eyeball that helps to focus the light source.

■ **Fovea** Centre of the macula, packed with cone cells.

■ **Geniculate body** Relay station to the optic radiation.

■ **Hyaloid canal** Remnant of the foetal blood vessel that supplied the lens.

■ **Iris** Pigmented continuation of the choroid coat.

■ **Lacrimal canaliculi** Ducts that run from the inner canthus to the lacrimal sac.

■ **Lacrimal gland** Situated above the other canthus.

■ **Lacrimal sac** Situated beside the inner canthus to collect fluid from the lacrimal canaliculi and drain through the naso-lacrimal duct.

## Anatomy of the eye

Labels: Sclera, Choroid, Retina, Fovea, Hyaloid canal, Blind spot, Retinal vessels, Optic nerve, Ocular muscle, Suspensory ligament, Cornea, Iris, Pupil, Aqueous humour, Lens, Ciliary body, Conjunctiva, Vitreous humour

## The eyeball

The eyeball (*above*) is nearly spherical and more than 2 centimetres in diameter. It is a globular structure filled with a jelly-like fluid under slight pressure to give it firmness. The white outer surface – sclera – surrounds the eyeball except for the transparent cornea in the front.

The conjunctiva covers the outer surface of the cornea as it bulges forward. The aqueous humour is produced by the ciliary body and circulates through the posterior, behind the iris, and anterior chambers of the eye to bathe the inner surface of the cornea and lens.

The lens is a soft, bi-convex, transparent structure in a thin, tough capsule. It divides the anterior third of the eye from the posterior two-thirds, being held by the suspensory ligament to the ciliary muscle fibres.

The ciliary body contains the ciliary muscles, which alter the shape of the lens. It is close to the ducts that change the aqueous humour and, together with the iris and the choroid coat, forms the uveal tract. Much of the focusing of the eye is made by the convex shape of the cornea; finer adjustments are made by the contraction of the ciliary muscles altering the lens shape.

The iris is the continuation of the choroid in front of the lens. Its colour, genetically determined, depends on the way in which the pigments are distributed. The pupil – the circular opening in front of the lens – varies in size very rapidly depending on the amount of light falling on the retina. The circular and radial muscle fibres in the iris are under the control of the autonomic nervous system. This prevents overstimulation of the retina by brilliant light. The pupil can vary in size from 1 to 8 millimetres. The vitreous humour fills the posterior two-thirds of the eyeball. Through the centre there is a thin, vessel structure that is the empty remnant of the foetal blood vessel that used to supply the lens in the foetus – the hyaloid canal.

The choroid coat lines the inner surface of the sclera and has brown pigmented cells to absorb light.

The optic nerve, containing some 1 million nerve fibres, penetrates the sclera and choroid coats and the nerves spread round the inner surface of the eyeball to form the retina. The point at which the optic nerve enters is known as the "blind spot" as there are no light-sensitive nerve cells at this point. The optic nerve is accompanied by an artery and vein that spread over the retina.

The retina consists of light-sensitive cells – the cones for red, green and blue, and the rods for shades from grey to white. The macula is the point at the back of the eye where the light will naturally fall when the eye is at rest. It has the greatest concentration of cones in the eye. The fovea is in the centre of the macula and is densely packed with colour-sensitive cones. It is responsible for the sharpest vision.

Visual acuity is the accuracy with which an object is seen; theoretically, the object must be of a size sufficient to stimulate one rod or cone. The two eyes work together – binocular vision – to relay the visual information to the visual cortex of the brain where the visual stimulus is interpreted in three dimensions.

## Eye muscles – left eye

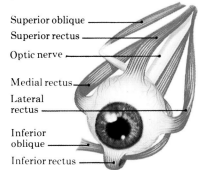

Superior oblique
Superior rectus
Optic nerve
Medial rectus
Lateral rectus
Inferior oblique
Inferior rectus

## Eye movements

There are six muscles to control the movements of each eye (*above*); the movements are co-ordinated in the brain. If the lateral rectus muscle in one eye contracts, the medial rectus of the other will contract to a similar extent. The superior recti work together to pull the eyes back and to look up. The inferior recti make the eyes look down. The superior oblique muscle rotates the eye downwards and outwards and the inferior oblique upwards and outwards.

## Pupil reflex

## Pupil size and light intensity

The retina is very sensitive to light (*above*). Too much light (**A**) distorts colour and is dazzling. The pupils vary in size and thus reduce or increase the amount of light entering the eye. Bright light causes a reflex nervous reaction, controlled by centres in the midbrain. The circular pupillary muscle (**1**) in both irises contracts and the radial fibres (**2**) extend, thus narrowing the diameter. Poor light (**B**) will make both pupils dilate, allowing sufficient light to stimulate the cells in the retina (**3**).

## The protective shield

The eyeball is deep set in the skull, under the orbital plate of the frontal bone and resting on the maxilla with the zygoma on the lateral side.

The hairy eyebrows, on the orbital ridge of the frontal bone, drain sweat and dirt to the side of the face. The long eyelashes of the upper lid prevent dust entering the eye by the automatic blinking reflex.

The upper eyelid contains a fibrous, tarsal plate that protects the closed eye when it firmly joins the softer lower lid. The upper lid is opened by the levator palpebra muscle and closed by the orbicularis oculi eye muscle, which also surrounds the lower lid. The edges of the lid are lined with modified sebaceous glands – the Meibomian glands – that keep the edges of the eyelid moist.

The conjunctiva is a thin, transparent membrane that lines both the inner surface of the eyelids and the outer surface of the eyeball; it acts as a protective layer and its smooth surface also allows easy movement of the eyelids.

## The lacrimal apparatus

The lacrimal gland lies within the orbit just under the frontal bone as it joins the zygoma. This is at the lateral canthus. It secretes a salty fluid that lubricates the conjunctiva. This is drained by lacrimal ducts, from the inner corner of each eyelid at the medial canthus, into a dilated lacrimal sac. This in turn drains into the naso-lacrimal duct running down into the nose.

The lacrimal secretion has three main functions: To keep the conjunctiva moist; to act as a mild antiseptic; and to drain away dust and any small particles that may have landed on the conjunctiva.

Irritants, such as dust or infection, and emotion will cause a rapid increase in lacrimal secretion to such an extent that the drainage system is flooded and tears will overflow down the face. Blinking is a normal, reflex action occurring several times a minute to help lubricate the eye. It also occurs as a reflex, protective mechanism to sudden movement close to the face.

## The retina

Light reaching the retina (*below*) has to pass through the nerve fibres – the ganglia (**1**) – joining them to the bipolar cells (**2**) that act as interconnecting neurons for the light-sensitive cells lying on the pigmented, retinal coat (**3**). The colour-sensitive cone cells (**4**, orange) are most densely congregated at the fovea (**5**). The rods (**6**, green) are found throughout the rest of the retina, giving wide black and white range.

## Rods

There are about 130 million rods and cones (photographed in detail below) in the ratio of 18 rods to 1 cone. The rods are more sensitive to light than the cones. There is a loss of colour vision and detail in poor light as there are fewer rods in the macula and none in the fovea. Light-sensitive, pigmented chemicals in the cones react to particular wavelengths of light – red, green or blue – giving a sensation of colour. The rods contain rhodopsin – visual purple.

## Rods and cones

Cone
Rod

Photo: Gene Cox

# Vision

■ **Retina** Layer of light-sensitive cells lying on the choroid coat of the eyeball.

■ **Rhodopsin** Light-sensitive pigment in the rods.

■ **Rods** Cells that respond to light but are unable to differentiate colour.

■ **Sclera** Tough, white outer coat of the eye.

■ **Suspensory ligament** Ligament that suspends the lens from the ciliary muscle.

■ **Tarsal plate** Thin lager of fibro-cartilage in the upper eyelid.

■ **Uveal tract** Consists of ciliary body, iris and choroid coat.

■ **Visual acuity** The accuracy of vision.

■ **Visual purple** *See* Rhodopsin.

■ **Vitamin A** Combines with enzymes to help reconstruct the pigments that are broken down by light.

■ **Vitreous humor** Semi-fluid gel behind the lens.

**Common diseases**

■ **Amblyopia** Dimness of vision.

■ **Astigmatism** Disorder of focusing that disturbs the visual shapes of objects.

■ **Blepharitis** Infection of the edges of the eyelids.

■ **Cataract** Opacity of the lens.

■ **Chalazion** Small, round swelling of one of the lubricating meibomian glands in the edge of the eyelid which is caused by blockage of the duct in the glands.

■ **Choroiditis** Inflammation of the choroid coat.

■ **Conjunctivitis** Infection of the conjunctiva.

■ **Diplopia** Double

■ **Glaucoma** Increased pressure within the eye that causes disturbance of vision and, if untreated, blindness.

**The visual mechanism**

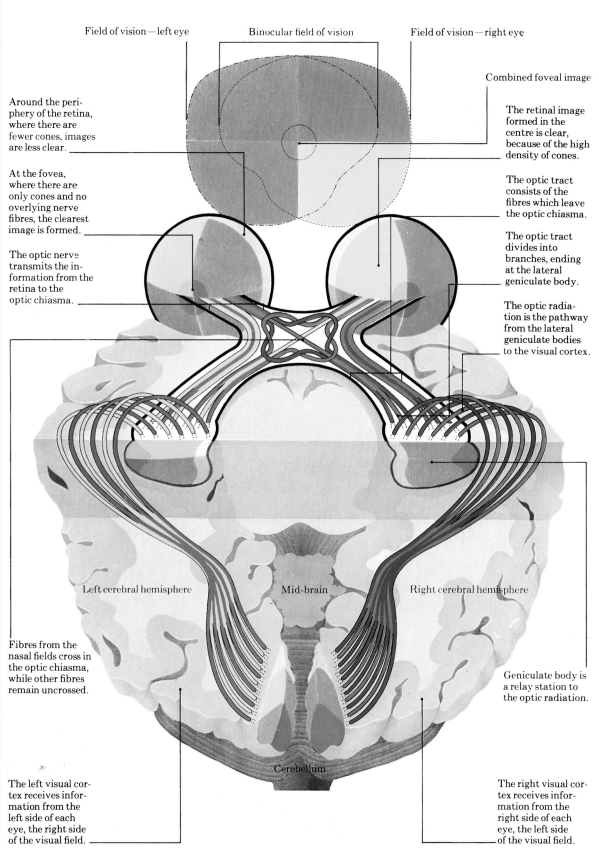

Field of vision—left eye

Binocular field of vision

Field of vision—right eye

Combined foveal image

Around the periphery of the retina, where there are fewer cones, images are less clear.

At the fovea, where there are only cones and no overlying nerve fibres, the clearest image is formed.

The optic nerve transmits the information from the retina to the optic chiasma.

The retinal image formed in the centre is clear, because of the high density of cones.

The optic tract consists of the fibres which leave the optic chiasma.

The optic tract divides into branches, ending at the lateral geniculate body.

The optic radiation is the pathway from the lateral geniculate bodies to the visual cortex.

Left cerebral hemisphere

Mid-brain

Right cerebral hemisphere

Fibres from the nasal fields cross in the optic chiasma, while other fibres remain uncrossed.

Geniculate body is a relay station to the optic radiation.

Cerebellum

The left visual cortex receives information from the left side of each eye, the right side of the visual field.

The right visual cortex receives information from the right side of each eye, the left side of the visual field.

66

## Transmission to the brain

The cross-section of a brain (*left*), viewed from above, shows how the image that reaches the retina is coded and relayed to the visual cortex. Stimulation of the retina gives an image of various colours, shades and shapes to be transmitted in the optic nerve to reach the optic chiasma. At the chiasma half the neurons in each optic nerve change sides. This allows the brain to interpret the pictures from each eye in the correct way; the image on the outer half of one eye is the same as the nasal half of the other eye. The nerve fibres now continue in the optic tract to the geniculate body, which helps to co-ordinate the eye movements with the midbrain, cerebellum and the autonomic nervous system. The retinal image is then sent on through the optic radiation to the posterior lobe of the cerebral cortex, where, for the first time, the visual stimulus is interpreted in the consciousness in three dimensions. It is linked with other lobes by the association tracts.

## Focusing mechanism

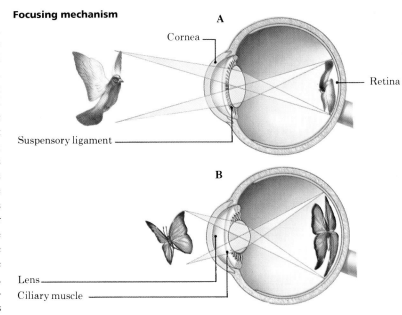

## Focusing mechanism

Focusing (*above*) is mainly dependent on the cornea and the relaxing and contracting effect of the ciliary muscles as they pull on the suspensory ligament of the lens causing fine focusing. In distant vision (**A**) the muscles relax and the ligaments pull the lens into a disc shape. Close vision (**B**) requires a more circular lens, so the muscles constrict and the ligaments relax.

## Colour blindness

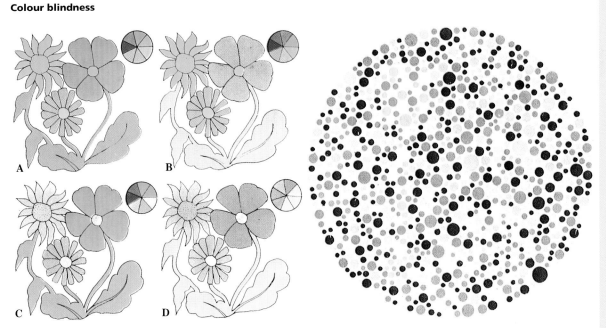

## Colour blindness

Colour vision (*above left*) in its normal form (**A**) is thought to depend on the ability of three different types of cones to respond to red, green and blue light wavelengths.

About 10 per cent of all men and less than 1 per cent of all women have genetically inherited defective colour vision. This is commonly red or green (**B**) or, rarely, blue (**C**). Total colour blindness (**D**) can occur.

Defects in colour vision can be detected using Ishihara test plates, like the one above right. The spade is purple and thus stimulates blue and green receptors, and cannot be seen by the green colour blind. The fork is red so it is not seen by the red colour blind. Both are seen by those with normal colour vision.

# Hearing

## The outer ear

In the ear (*right*) the outer ear consists of the pinna and external auditory meatus. The pinna is a thin sheet of cartilage moulded to direct sound towards the auditory meatus. It is covered with skin and, in some people, can be moved by weak muscles attached to the skull. The external auditory meatus is a blind tube 2.5 centimetres long ending at the tympanic membrane. In the outer third it is composed of cartilage and the skin is covered with protective hairs. The inner part lies in the temporal bone and contains modified sweat glands – ceruminous glands – that produce a waxy secretion – cerumen – to protect the skin and to drain out dust and bacteria.

## The middle ear

The cavity of the middle ear lies deep in the temporal bone. The fibrous tympanic membrane stretches tightly across the end of the external auditory meatus. It is ovoid in shape, measuring about 9 millimetres across at its widest point, and transmits sound to the three small interlocking bones. The malleus is attached to the membrane and the incus links it with the stapes by two synovial joints. The stapes is attached to the foramen ovale of the inner ear. These bones vibrate in unison and magnify sound twenty times. The minute stapedius muscle, supplied by the VII cranial nerve, is attached to the stapes; the tensor tympani muscle, supplied by the V cranial nerve, tightens the tympanic membrane by its attachment to the malleus. Both muscles dampen excessive movement.

The air pressure in the middle ear is kept at atmospheric levels by the Eustachian tube, which connects with the posterior nasal space. The mastoid air cells, in the temporal bone, open into the posterior part of the middle ear.

## The inner ear

The inner ear contains the cochlea, semicircular canals and auditory nerve.

The two ears, working in unison, can locate the source of a sound by the difference in volume and timing as it reaches the two sides of the head.

**The ear**

Pinna

Semicircular canals

Vestibular branch of VIII nerve

Malleus

Tympanic membrane

Auditory branch of VIII nerve

Cochlea

Incus

Stapes

Utricle

Saccule

VIII nerve

External auditory meatus

Eustachian tube

**The organs of balance**

Semicircular canals

Utricule

Saccule

Ampulla

Receptor cells

Receptor cells with otoliths

## The organs of balance

The organs of balance lie deep in the petrous part of the temporal bone on either side of the skull adjacent to the cochlea in the inner ear (*above*). The three semicircular canals are at right angles to each other so any movement of the body will cause movement of the endolymph to stimulate the hair-like fibres, supplied with nerves, in the ampullae sited at the base of the canal.

The endolymph will continue to move after rapid movement even when the body itself is stationary. The continued movement of the fine hairs caused by the fluid indicates to the brain that the head is moving, but the eyes tell it that movement has stopped. These conflicting signals give the sensation of vertigo.

Interconnecting with the ampullae of the three canals are the utricle and the saccule. They also contain hair cells with otoliths – minute crystals of calcium carbonate suspended in a jelly-like material. Pressure from the weight of these crystals stimulates the nerve endings when the head is stationary. Standing and lying will put pressure on different receptors.

The vestibular branch of the VIII cranial nerve supplies the semicircular canals, utricle and saccule. Information is taken to the cerebellum, which co-ordinates movement and balance with the long sensory and motor tracts in the brain stem via the pons.

## Organs of hearing

The two cochlea also lie in the petrous part of the temporal bones. Sound vibrations reach them through the foramen ovale from the stapes bones. Each is a narrow cone-shaped tube, 2.5 centimetres long, coiled two and a half times (*right*). They have three compartments: Scala vestibuli; tympani; and media.

Fluid vibrations of the perilymph pass up the scala vestibuli to the top of the cochlea and then down the tympani to reach the foramen rotunda in the medial wall of the middle ear. This allows the fluid to move by bulging when the pressure increases.

The human ear can detect sound between about 20 and 20,000 cycles per second (Hertz). The intensity of sound is measured in decibels (dB). A whispered voice is about 10 dB, a jet plane at take-off 110 dB. Decibels are a logarithmic measurement, so the difference between 40 and 50 dB is twice the volume, not 25 per cent as it might seem. Sound above 100 dB is damaging to the ear and it may lead to damage and deafness.

### The cochlea

Scala media
Organ of Corti
Scala vestibuli
Scala tympani
Sound waves
Auditory branch of VIII nerve

### The organ of Corti

In the scala media, which runs the full length of the cochlea and contains endolymph, lies the organ of Corti (*below*). It has five rows of cells on a basilar membrane, each with neat rows of hair-like projections. These cells are held in place by special supporting cells. The tips of the hairs are inserted into the gelatinous tectorial membrane, which overhangs them. Sound waves in the perilymph of the scala vestibuli and tympani cause the basilar membrane to vibrate at the appropriate wave-length. This stimulates the hair cells, which vibrate and activate one of the 30,000 neurons of the auditory part of the VIII cranial nerve.

Sounds with more than one frequency will stimulate several areas of the organ of Corti. High frequencies activate the first part of the basilar membrane. The auditory nerve goes to the brain stem to reach the cochlear nucleus. Some fibres cross to the other side of the brain and interconnect with other nuclei; the rest pass to the auditory centre in the temporal lobe.

■ **Tympanic membrane** Ear drum.

■ **Utricle** With the saccule forms part of the organ of balance.

■ **Vestibular nerve** Part of the VIII cranial nerve associated with balance.

**Common diseases**

■ **Deafness** – Interference with sound reaching the cochlea owing to damage to the cochlea or sensory nerve or failure of conduction.

■ **Ménière's disease** Chronic condition due to intermittent raising of pressure of the endolymph of the ear.

■ **Otitis externa** Inflammation of external auditory meatus.

■ **Otitis media** Infection of the middle ear.

### The cochlea

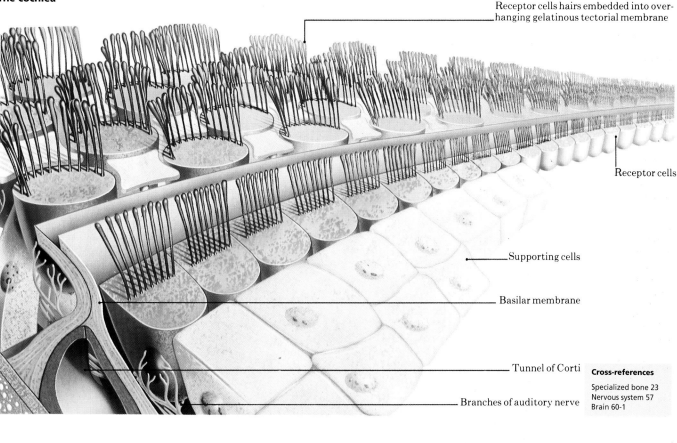

Receptor cells hairs embedded into overhanging gelatinous tectorial membrane

Receptor cells

Supporting cells

Basilar membrane

Tunnel of Corti

Branches of auditory nerve

# Skin, Hair and Nails

70

## Definitions

■ **Albino** Congenital absence of pigment-forming enzyme.

■ **Apocrine gland** Particular kind of sweat gland that is found in the axillae and pubic areas producing a thick form of sweat.

■ **Areola** Circular area around nipple.

■ **Ceruminous gland** Modified sweat gland that is found in the ears and produces wax-like cerumen.

■ **Dermis** Inner layers of skin lying beneath the epidermis.

■ **Eccrine gland** Sweat gland.

■ **Epidermis** Outer, cellular layer of skin that forms keratin.

■ **Erector muscle** Small muscle attached to hairs in the dermis.

■ **Follicle** Cells forming hair.

■ **Hair** Fibre of dead keratinized cells.

■ **Keratin** Hard, dead, horn-like material.

■ **Lactiferous sinus** Tube into which the milk is secreted.

■ **Melanin** Dark pigment.

■ **Melanocyte** Cell that produces pigment to be incorporated in other cells.

■ **Sebaceous gland** Grease-producing gland in the skin.

■ **Sweat glands** Glands that produce highly dilute solutions of various salts and urea; they help in the temperature regulation of the body.

## Common diseases

■ **Acne vulgaris** Blockage of the sebaceous gland by a mixture of sebum, keratin and dead cells, which forms a "blackhead".

■ **Boil** Staphylococcal skin infection of hair follicle with the formation of pus.

■ **Carbuncle** More severe form of boil with several points of discharge.

### The skin – the protective coat

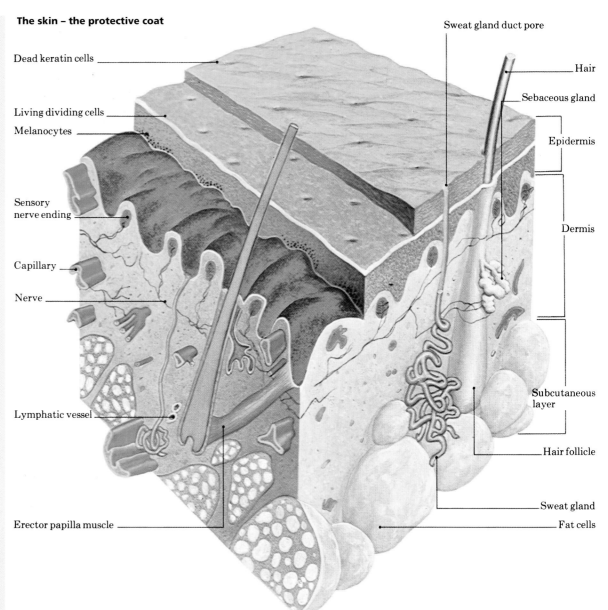

Dead keratin cells

Living dividing cells

Melanocytes

Sensory nerve ending

Capillary

Nerve

Lymphatic vessel

Erector papilla muscle

Sweat gland duct pore

Hair

Sebaceous gland

Epidermis

Dermis

Subcutaneous layer

Hair follicle

Sweat gland

Fat cells

## The skin

The skin (*shown in detail above*) covers an area of about 1.7 square metres and weighs approximately 3 kilogrammes. It is waterproof and a defence against trauma and infection; it is also a sensory organ and a temperature regulator. The outer layer of the epidermis is of dead, keratinized cells, which flake away all the time. Beneath this the epidermal living cells are growing outwards as they gradually form the keratin.

Underneath is the dermis, consisting of a network of collagen tissue fibres with interweaving blood and lymph vessels, sweat and sebaceous glands, and hair follicles and nerve endings; the latter are adapted to detect pain or pressure, heat or cold and vibration. Below the dermis is a variable layer of fat-storage cells.

The skin colour depends on the degree of melanin. All races have the same number of pigment cells – melanocytes – but genetic differences control the amount incorporated in the epidermal cells. Sunlight stimulates further production. Albinism is due to an absence of pigment-forming enzyme. About 3 million eccrine sweat glands in the skin produce water, waste salts and urea. In the ear they are modified into ceruminous glands to produce cerumen – wax. A different type of sweat gland – apocrine – occurs in the axillae and pubic areas. These secrete a thicker fluid that the skin bacteria breaks down to give body odour. Sebaceous glands, adjacent to the hair follicles, produce sebum – a fatty secretion that oils the hair and lubricates the skin by keeping the sweat on the epidermis.

## The breast

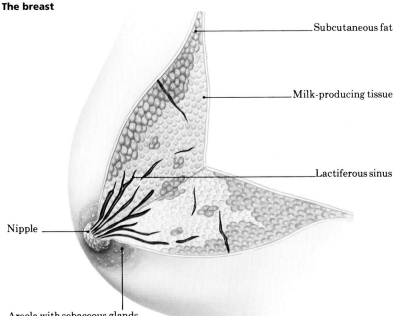

Subcutaneous fat

Milk-producing tissue

Lactiferous sinus

Nipple

Areola with sebaceous glands

## The nail

Lunular pale crescent of active nail growth

Nail formed of keratin

Bone—one of the distal phalanges

Cuticle fold of epithelial cells

Nail bed formed from living layer of epidermis

## Hair

The hair follicle in the dermis is a tube with a central core of dead, keratinized cells, which are pushed upwards as new cells are produced to form a hair. Each follicle has a small erector muscle that can contract to make the hair stand upright. Only the palms of the hands and soles of the feet have no hair.

There are one to 200,000 hairs on the scalp which grow about 2 millimetres a week. The colour of the hair is dependent on its melanin content; the hair becomes whiter with age due to less pigmentation and minute air bubbles in the shaft. Balding, in part, is due to the male hormone – testosterone – produced in the testes.

## The breast

The breast is composed of 15 to 20 modified sweat glands which develop into lobes. They are rudimentary in the male. The female breast gland (*top left*) develops rapidly at puberty with secreting cells responding to the hormones in the menstrual cycle. During pregnancy the glands become congested and milk is collected in the lactiferous sinuses, which join behind the areola of the nipple. The areola is lubricated by the moist secretion of sebaceous glands.

## The nails

The nails (*centre left*) develop at the third month of foetal life from a special layer of cells that fills with hard keratin and grows over the epidermis at the ends of the fingers and toes as a protective layer.

## Temperature control

The temperature centre in the hypothalamus controls heat loss and production by the body through the skin (*below*). Overheating (**A**) causes an increased blood flow, from the blood vessels (**1**), to radiate heat and cause sweating, through the sweat glands (**2**), to lose heat. A fall in temperature (**B**) constricts the surface blood vessels, stops sweating and makes the erector muscles (**3**) contract, causing the hairs (**4**) to stand on end, trapping air as an insulating layer. Additional heat can be produced by shivering.

## Temperature control

■ **Dermatitis** Inflammatory reaction of the skin that may be due to infection, contact with an allergy – producing substance or as a reaction to some internal disease.

■ **Eczema** Red, rough skin usually taken to refer to atopic eczema, an inherited condition in which there is roughness of the skin, particularly in the flexures of the arms and legs, but sometimes the whole skin.

■ **Keloid scar** Overgrowth of the healing fibrous tissue of the scar that causes a hard, slightly raised, reddened surface.

■ **Moles** Raised, slightly pigmented areas of the skin found in most people. Only very rarely do they become malignant.

■ **Paronychia** Infection at the nail bed.

■ **Pediculosis** Infestation of the scalp or pubis with lice.

■ **Prickly heat** Blockage of the sweat glands occurring after prolonged sweating in a hot climate.

■ **Seborrhoeic dermatitis** Scaling redness of the scalp, eyebrows, axillae and the face. It is due to an inborn tendency and may be aggravated by mild infection.

■ **Tinea** Fungal infection of skin.

■ **Verruca** Viral wart.

**Cross-references**

Muscles 27
Brain 60–1
Pregnancy 80–5

# Touch, Taste and Smell

## Skin sensation

Some areas of the skin are densely packed with nerve endings, as in the finger-tips, while others, as on the back, have comparatively few. This is reflected in the proportion of the parietal cortex devoted to various areas of the body.

The nerve endings which are concerned with touch, pressure and slight variations in temperature have similar structures. Slight differences in both structure and position have led to the nomenclature illustrated (*right*). Nerve endings in the hair follicle detect hair movement.

Many nerve endings do not have end organs and these detect light, touch and pain. Sufficient stimulus will cause pain in any nerve ending. The brain correlates all skin sensations with the eyes, ears and nose

**Skin sensation**

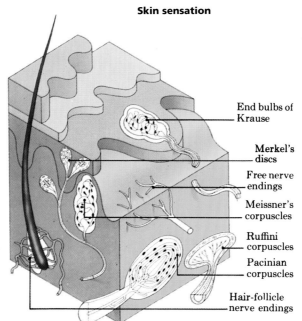

End bulbs of Krause

Merkel's discs

Free nerve endings

Meissner's corpuscles

Ruffini corpuscles

Pacinian corpuscles

Hair-follicle nerve endings

## Referred pain

Internal organs and structures are well supplied with nerves, but pain is diffuse and poorly located compared with skin sensation (*below*). Most of the pain is produced by stretching and contracting, hence the pain of colic. Internal pain will cause stimulation of local nerves in a segment of the spinal cord; this makes it appear that the pain is coming from the skin, which is supplied by the sensory nerves. The heart (**1**) and oesophagus (**2**) refer pain to the neck, shoulders and arms; uterus (**3**) and pancreas (**4**) to the lumbar region; and kidneys (**5**) into the groin. Diaphragmatic pain may be referred to the shoulders, as the phrenic nerve of the diaphragm is formed from the spinal nerves in the neck, which also supply the shoulders.

**Smell – the nose**

**Referred pain**

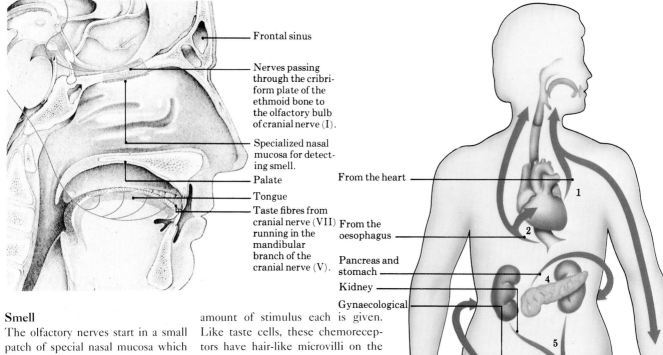

Frontal sinus

Nerves passing through the cribriform plate of the ethmoid bone to the olfactory bulb of cranial nerve (I).

Specialized nasal mucosa for detecting smell.

Palate

Tongue

Taste fibres from cranial nerve (VII) running in the mandibular branch of the cranial nerve (V).

From the heart

From the oesophagus

Pancreas and stomach

Kidney

Gynaecological

## Smell

The olfactory nerves start in a small patch of special nasal mucosa which is found high in the nasal passages (*above*). They pass through minute holes in the cribriform plate of the ethmoid bone into the olfactory bulb of the cranial nerve (I) before entering directly into the limbic system of the brain.

It is thought that there are several types of chemoreceptors and that odours are differentiated by the amount of stimulus each is given. Like taste cells, these chemoreceptors have hair-like microvilli on the surface. The dissolved, odiferous molecules may fit into some, but not into all, types of cell microvilli, and so causing an electrical discharge and resulting stimulus of the nerve. There is certain evidence to suggest that some substances, pheromones, can act as sexual stimulants. Well-trained sense receptors can recognize over 10,000 different odours.

# Taste—the tongue

Epiglottis

The base of the tongue is devoid of papillae, but is covered with nodules of lymphoid tissue, which go to make up the lingual tonsil.

Filiform papillae

Fungiform papillae

A cross-section of a vallate papilla, showing the moat in which dissolved food bathes the taste bud.

A taste bud consists of taste cells, with hairs that project into the moat, supporting cells and nerves.

The salivary glands are connected by nerves to the taste buds.

A cross-section of a salivary gland showing drops of saliva, produced by the salivary cells, lying in the duct of the gland

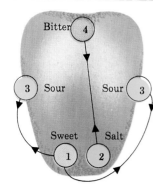

Bitter 4

3 Sour    Sour 3

Sweet    Salt

1    2

## Taste

The highly mobile muscular tongue (*above*), supplied by the hypoglossal nerve (**XII**), is attached to the hyoid bone and membrane between the mandibles. It aids mastication and articulation of speech. The surface is covered by thick epithelium containing about 9,000 papillae – taste buds – which are re-formed and replaced within 48 hours, but become fewer with age.

There are four kinds of papillae: Filiform, folioform and vallate, which are only found at the back of the tongue, and fungiform. Each contains one to two hundred microvilli covered with chemoreceptors. A single nerve may connect several cells.

All flavours depend on a mixture of four basic tastes (*see* inset): Sweet (**1**) and salt (**2**), mainly on the front, and sour (**3**), at the sides, and bitter (**4**) at the back.

The glossopharangeal nerve (**IX**) carries sensation from the posterior third of the tongue. Taste nerves in the mandibular branch of the nerve (**V**), which supply the rest of the tongue, branch to join the facial nerve (**VII**).

Palatability is a combination of taste and temperature, and feel and smell. It is assessed by the parietal lobe of the brain in conjunction with the thalamic taste centre.

# Female Reproductive Organs

**Definitions**

■ **Acrosome cap**
Layer of tissue on the head of the sperm. It disappears when it dissolves the ovum's membrane.

■ **Bartholin's gland**
One of a pair of lubricating glands sited adjacent to the posterior entrance of the vagina.

■ **Cervix** Muscular structure, the lower part of the uterus, with a central canal connecting the vagina to the uterine cavity.

■ **Clitoris** Small organ, composed of erectile tissue, that lies anterior to the entrance of the vagina.

■ **Corpora cavernosa** One of a pair of erectile tissue organs that overlie the urethra and the corpus spongiosum.

■ **Corpus albicans**
Small, white, scarred area left by the disintegrating corpus luteum.

■ **Corpus luteum**
Cellular structure that forms from the follicular cyst after ovulation; producing progesterone and oestrogen.

■ **Corpus spongiosum** Tube of erectile tissue that surrounds the penile portion of the urethra and ends in the glans of the penis.

■ **Cowper's gland**
Pair of glands, adjacent to the prostate gland, whose secretions help with the nutrition of the sperm in the ejaculate.

■ **Ectoderm** Layer of cells in the inner cell mass that eventually develops into the skin and nervous system.

■ **Ejaculate** Seminal fluid, secretions of Cowper's and Littre's glands, that is produced by the male during intercourse.

■ **Endoderm** Layer of cells of the inner cell mass that forms the alimentary canal, liver, trachea and lungs.

**Female reproductive organs**

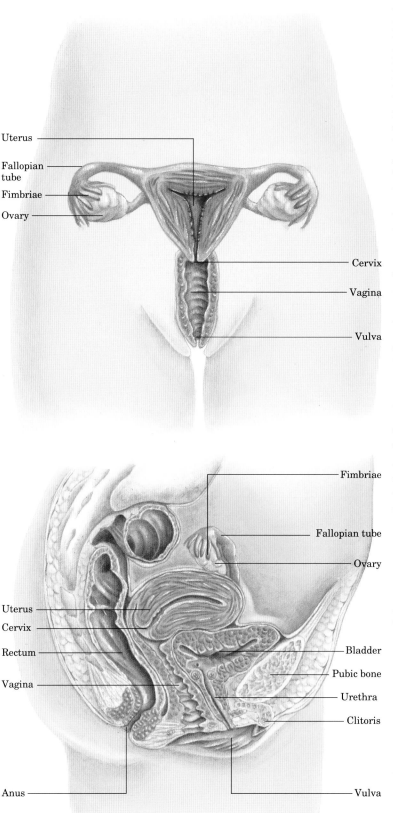

The female reproductive system not only has to produce an ovum but has to nurture the fertilized ovum and protect it until pregnancy ends. At the entrance to the vagina a pair of lip-like folds – the larger, thicker labia majora and the smaller, inner labia minora – lie along either side of the vaginal entrance and join in the front, blending into the padded, hairy area of the mons pubis. At the front they enclose the exit of the urethra just behind the small projection of erectile tissue – the clitoris – comparable to the penis.

Behind these structures is the vagina – a 10 to 15 centimetre elastic tube lined with moist epithelium; it is normally "closed" except during intercourse. At the back of the vulva are two large Bartholin's glands, and lining the entrance to the vagina a large number of smaller, lubricating glands. At the top of the vagina the uterus is held in place by the muscles and four strong fibrous ligaments of the pelvic floor, and to the side of the pelvis by pairs of the round and suspensory ligaments, running in folds of peritoneum.

## The uterus

The uterus is a small pear-shaped organ, covered with peritoneum, with a thick wall of interweaving muscle fibres and lined with the special endometrial cells. It is situated behind the bladder and in front of the rectum. The cervix of the uterus is a thick fibro-muscular structure opening into the vagina and lined with special cells that form a "plug" of mucus. The uterine muscle is always contracting and relaxing slightly. This is increased during sexual orgasm in order to "suck in" sperm; during menstruation, to expel the endometrium; and also at parturition.

The two Fallopian tubes are about 10 centimetres long with finger-like fimbriae at the ends to encircle the ovaries. The ovum is swept down the tube by a combination of ciliated epithelium and peristaltic muscular contraction. Unlike the male urinary system, that of the female is separate from the reproductive system. The bladder empties into the urethra, which opens in front of the vagina.

## The ovary

The ovary has two functions: The production of ova (*right*); and the secretion of oestrogen and progesterone. It is about 2 centimetres across and 1 centimetre thick.

At puberty, the onset of menstruation and the development of the secondary sexual characteristics – hair growth, breast development and redistribution of fat to the buttocks and shoulders – are the result of the effect on the ovaries of the increasing pituitary secretion of follicular stimulating hormones (FSH) and luteinizing hormones (LH).

Like spermatogenesis an oocyte is formed from the oogonia. Meiotic division then occurs to produce a large secondary oocyte with most of the cytoplasm from its "twin" – the polar body. Further mitotic division occurs so that one primary oocyte has produced a large ovum and three polar bodies, which finally disintegrate in the ovum.

The ovary has 50 to 250,000 oogonia, but only about 500 eventually become mature ova.

## The menstrual cycle

From menarche to the menopause the anterior pituitary lobe maintains a rhythm of FSH and LH secretion to produce, in most women, a regular menstrual cycle (*right*). FSH stimulates the maturation of several Graafian follicles, of which only one reaches maturity. The follicular cells produce oestrogen to build up the endometrium. In mid-cycle a surge of LH softens the plug of cervical mucus and causes ovulation. The corpus luteum is formed and secretes progesterone, which, with the oestrogen, further prepares the endometrium for implantation of the fertilized egg.

If fertilization takes place, the embryo produces human chorionic gonadotrophin to maintain the stimulus to the corpus luteum and continued production of progesterone. If fertilization does not occur the corpus luteum degenerates, progesterone production ceases and the endometrium is shed causing menstruation for about five days. The change in hormonal balance may be responsible for premenstrual tension.

### The ovary

### The ovary

The cross-section of the ovary (*above*) shows an immature ovum (**1**) absorbing fluid and swelling into the sac-like Graafian follicle (**2**). As it reaches the surface (**3**), half-way through the menstrual cycle, the ovum bursts out into the peritoneal cavity to be collected by the fimbriae.

The empty follicular cyst then changes into the corpus luteum (**4**), which grows until menstruation, when it shrivels into a scab-like corpus albicans (**5**).

### The menstrual cycle

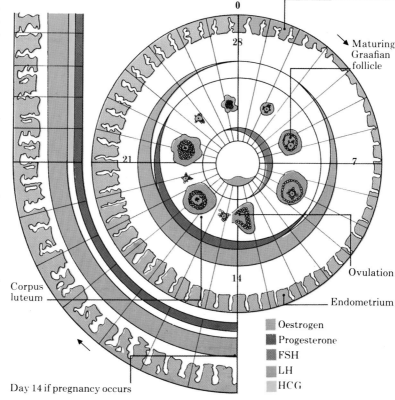

Day 14 if pregnancy occurs

Oestrogen
Progesterone
FSH
LH
HCG

■**Graafian follicle**
*See* Follicular cyst.

■**Endometrium**
Hormone-responsive cells that line the uterine cavity.

■**Epididymis**
Tightly coiled 6-m tube joined to the upper part of the testis by 12 to 20 vasa differentia, in which the sperm spend 10 days maturing .

■**Fallopian tube**
Two l0-cm tubes extending from the fimbriae, which join the body of the uterus.

■**Fimbria** One of the soft finger-like processes at the peripheral end of the Fallopian tube.

■**Follicular cyst**
Ovarian cyst that develops each month and from which the mature ovum bursts into the peritoneal cavity.

■**Glans penis**
Swelling at the distal end of the corpus spongiosum, which is covered by the foreskin in the uncircumcised male.

■ **Graafian follicle**
*See* Follicular cyst.

■ **Implantation**
Process by which the embryo erodes the endometrial lining of the uterus.

■**Labia majora** Two larger and outer folds of skin that surround the entrance to the vagina.

■**Labia minora**
Two smaller and inner folds of skin that surround the entrance to the vagina.

■**Lucuna** Small areas in the endometrium formed by the trophoblast, in which the maternal blood bathes the trophoblastic cells

■**Littre's gland** One of a pair of glands adjacent to the prostate whose secretion help maintain the health of the sperm in the ejaculate.

■**Mitochondria**
Minute structures within the cell that store energy required for cell metabolism.

# Male Reproductive Organs

**Male reproductive organs**

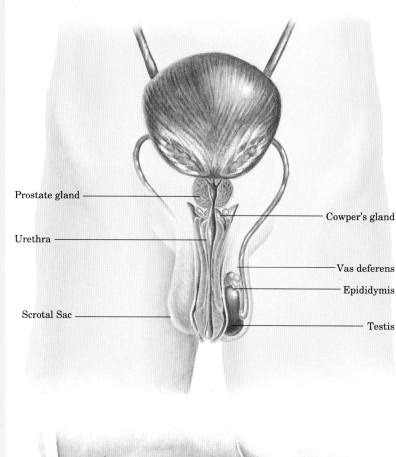

Prostate gland

Urethra

Scrotal Sac

Cowper's gland

Vas deferens

Epididymis

Testis

Bladder (full)

Prostate gland

Pubic bone

Urethra

Seminal vesicle

Rectum

Ejaculatory duct

Cowper's gland

Anus

Epididymis

Vas deferens

Testis

Corpora cavernosa

Corpus spongiosum

Much of the male reproductive anatomy is external. The two testes hang in the scrotal sacs surrounded by the tunica vaginalis. The vasa efferentia join the testes to the overlying epididymis and vas deferens, which join the urethra in the centre of the prostate gland. Akin to the gall bladder the seminal vesicle acts as a storage organ for the mature sperm. It lies between the prostate gland and the colon.

The wrinkled subcutaneous area of the scrotum contains the dartos muscle, which can contract and, with the cremaster muscle, attached to the spermatic cord, pull the testes closer to the abdomen. The spermatic cord is the combination of the vas deferens (surrounding testicular blood vessels and nerves), fatty tissue and the projection from the peritoneum that originally formed the tunica vaginalis.

### The prostate gland

The prostate gland lies around the first part of the urethra at the base of the bladder, and its secretions help maintain sperm activity. Two additional pairs of glands, Cowper's and Littre's, empty into the urethra adjacent to the prostate. They produce the bulk of the ejaculate of 3 to 5 millilitres of semen.

Semen contains various proteins, fructose and a mixture of chemicals, which help with the nutrition of sperm in the vagina. Each millilitre contains about 100 million sperm, of which 20 per cent are dead or abnormal. The number of sperm produced at ejaculation is variable.

The penis surrounds the fibroelastic urethra, which runs in the corpus spongiosum and ends in the glans, which is covered by the foreskin. The urethra and corpus spongiosum are covered, above and at the sides, by the pair of corpora cavernosa. Normally the penis is limp, but if the muscles at the base contract around the veins the spongy tissue in all three corpora becomes congested with blood and an erection occurs. The penis has two functions: The excretion of urine from the bladder out of the body; and the deposition of semen in the vagina of the female genital tract.

## The testis

The testis (*right*) has two functions: The production of testosterone; and of spermatozoa. It is about 5 centimetres long and 2.5 centimetres thick and surrounded by the tunica albuginea; this coat is divided into about two hundred lobes containing four to six hundred seminiferous tubules, each about 75 centimetres long. The tubules produce about 200 million sperm every day which pass into a series of communicating ducts – the rete testis. These ducts, lined with ciliated cells, join to form 12 to 20 vasa efferentia, which run to the epididymis, where the sperm spend up to ten days maturing before entering the vas deferens. The sperm can then survive a further six weeks before degeneration and absorption.

Spermatogenesis requires stimulus from the follicular stimulating hormone of the anterior pituitary gland. Testosterone responds to luteinizing hormone. It causes development of male secondary sexual characteristics – pubic and facial hair growth, aggressiveness, muscle bulk and deepening of the voice.

### The testis

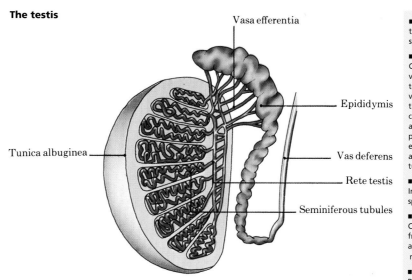

- Vasa efferentia
- Epididymis
- Vas deferens
- Rete testis
- Seminiferous tubules
- Tunica albuginea

## Spermatogenesis

The spermatogonia divides (*below*) by mitosis to form the primary spermatocyte, containing 46 chromosomes. Meiosis, on the other hand, occurs when half of each chromosome pair enters the secondary spermatocyte. This divides again in order to produce a spermatid that finally matures into a spermatozoon.

## The spermatozoa

The mature sperm (*below*) is 0.05 millimetres long. It consists of a head, body and tail. The head is covered by the acrosome cap and contains a nucleus of dense genetic material from the 23 chromosomes. It is attached by a neck to a body containing mitochondria that supply the energy for the sperm's activity. The tail is made of protein fibres that contract on alternate sides, giving a characteristic wave-like movement that drives the sperm through the seminal fluid, which also supplies additional energy. Some sperm have two heads or two tails and if the testes are too warm they may die or spermatogenesis may not occur.

### Spermatogenesis

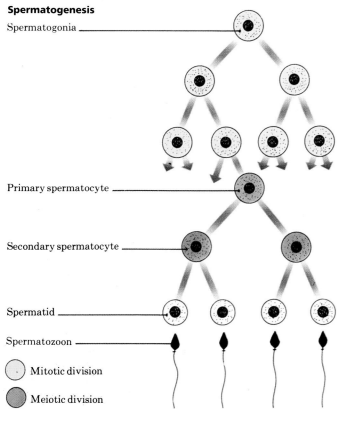

- Spermatogonia
- Primary spermatocyte
- Secondary spermatocyte
- Spermatid
- Spermatozoon

○ Mitotic division

● Meiotic division

### Spermatozoa

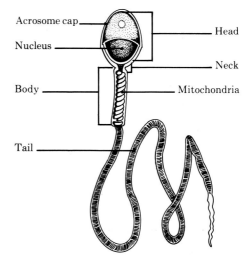

- Acrosome cap
- Nucleus
- Body
- Tail
- Head
- Neck
- Mitochondria

# Reproduction

## Coitus

Coitus is the insertion of the erect penis into the vagina followed by rhythmical movements ending in orgasm and ejaculation of semen by the man. Initially sexual excitement is produced by sight, touch and sound and, perhaps, pheromones that stimulate smell. In the man there is congestion of the corpora cavernosa and corpus spongiosum of the penis; in the woman there is slight engorgement of the breasts, and congestion of the clitoris and labia, with increased vaginal secretions. A prolonged phase of enjoyment can then follow when the penis is inserted into the vagina provided the "crescendo" to orgasm is controlled.

Orgasm is a series of rapid muscular contractions surrounding the male urethra that result in the ejaculation of the semen. This can be complemented by a simultaneous happening in the woman in which the upper part of the vagina and a rhythmical "sucking in" of the uterus draw the sperm to the right area. During sexual intercourse there are increases in blood pressure, rates of heartbeat and respiration. After orgasm there is physical relaxation.

## Fertilization

The moment of conception is the most important stage of sexual reproduction. The joining of the two nuclei, each containing 23 chromosomes, to form a cell of 46 chromosomes, will produce an embryo.

The sex of the baby is decided by the father's chromosomes – the XY sex pair. When the primary spermatocyte divides by meiosis to form the secondary spermatocytes, containing 23 chromosomes, either an X or a Y chromosome will move into each cell. The female pair of chromosomes are the same, XX, so all ova will contain the X chromosome. The fertilizing sperm will join the ova to form an XY male or XX female embryo. It is probable that there is less than 24 hours in which the ovum can be fertilized and as the sperm survive for about 48 hours in the uterus and tubes, there is only a limited time each month during which conception is able to take place.

**Sperm transport**

Although about one hundred sperm reach the ovum, only one sperm fuses with its nucleus.

Uterine contractions carry the sperm toward the Fallopian tube, but only one thousand enter.

Enzymes dissolve cervical mucus, allowing about one million sperm to enter the uterus.

In the vagina the mucus in the semen must be dissolved by enzymes before the sperm can move.

Up to a quarter of the 350 million sperm in one ejaculate are abnormal.

## The sperm's journey to the ovum

The ejaculate of semen contains more than 350 million sperm. Ideally they are deposited adjacent to the cervix, where the vaginal enzymes dissolve the seminal mucus and release the sperm. They then swim into the cervical mucus, which is watery and soft enough at ovulation to allow many of the sperm to penetrate to the uterine cavity. Uterine contractions will then help them to move upwards to the Fallopian tubes, where they again swim, perhaps with the help of the lining of ciliated cells, to reach the ovum at the mid-point of the tube. Only about 100 sperm survive the journey of nearly 24 hours and only one fertilizes the ovum. The ultimate fate of the surplus spermatozoa in the female genital tract is still not known.

## The moment of conception

Sperm

Acrosome

Ovum

The sperm's acrosome disappears as it dissolves the membrane of the ovum (*above*). The tail and body are shed when the head penetrates to join its 23 chromosomes with those of the ovarian nucleus.

## Implantation of the ovum

Blastocyst — Morula — Blastomere

**1**

**2**

**3**

Trophoblast

Implantation of blastocyst
**4**

Inner cell mass

Endometrium

**5**
Amniotic cavity

Ectoderm
Endoderm
Yolk sac
Lacunae

### Implantation of the ovum

It takes about a week for the fertilized ovum to pass down the Fallopian tube and implant itself in the endometrium. Within hours of conception mitosis begins with the development of a sphere of an increasing number of cells; the sphere starts as the blastomere (**1**), then becomes a morula (**2**) of about 64 cells. At this stage it changes into a hollow, fluid-containing ball – blastocyst (**3**) – with the inner cell mass at one end. It can now begin implantation (**4**) into the lining of the uterus.

The outer ring of cells, now called the trophoblast, secretes enzymes that erode the endometrium. The trophoblastic cells spread into the endometrium forming lacunae – fluid-containing sacs – that penetrate the maternal circulation. This allows nutrition to take place. Ultimately the trophoblast forms the outer layer of the placenta.

The inner cell mass splits into two layers: Endoderm – ultimately the alimentary tract – which produces a yolk sac; and the ectoderm – ultimately the skin, brain and spinal cord

– which produces the amniotic sac to surround and protect the embryo in a bag of fluid.

By the ninth day after conception the blastocyst has sunk deep into the endometrium (**5**) and is already receiving nutrition from the mother. Human chorionic gonadotrophin is secreted by the trophoblast to maintain the progesterone production from the corpus luteum. This will continue until the end of the third month, when the placenta produces sufficient hormone to maintain this function until the end of pregnancy.

■ **Bartholin's abcess** Infection of the Bartholin's gland.

■ **Cervical carcinoma** Most common in women who have had frequent sexual intercourse at an early age. The early stages can be detected by the cervical smear test.

■ **Cervical erosion** Change in the type of cells on the cervix that makes it more vulnerable to infection.

■ **Dysmenorrhoea** Painful periods.

■ **Endometritis** Infection of the uterus.

■ **Menorrhagia** Excessive menstrual bleeding.

■ **Ovarian carcinoma** Various forms of cancer that may occur in the ovary.

■ **Ovarian cyst** This may be an abnormally large follicular cyst or one that has developed independently.

■ **Salpingitis** Infection of the Fallopian tube.

■ **Vaginitis** Infection of the vagina.

**Common diseases**
(Male or female)

■ **Infertility** This may occur due to failure to ovulate or produce adequate numbers of sperm. Blockage of the Fallopian tube or vas deferens will prevent the sperm and ovum meeting.

■ **Venereal diseases** Diseases that are transmitted by sexual intercourse such as HIV infection gonorrhoea, syphilis and non-specific urethritis, which mainly afflicts males. There are other venereal diseases which are commoner in tropical countries.

**Cross-references**

Cell 17–19
Urinary system 50–2
Endocrine 53–5

# Pregnancy (Months 1 – 4)

**Development of cell layers**

Trophoblast

Yolk sac

Embryo with three types of cell

Amnion

**Mesoderm**
Bone
Cartilage
Lining of blood vessels
Lining of ureters
Muscle

**Endoderm**
Surface of alimentary canal and associated glands
Surface of lungs, trachea and larynx

**Ectoderm**
Brain
Epidermis
Hair
Lining of nose, mouth and anus
Nails
Nerves
Sensory organs

## The first four weeks

The blastocyst begins to implant into the endometrium on the seventh day after conception. It is composed of two distinct parts, a covering layer of the trophoblastic cells and an inner cell mass.

The trophoblast invades the maternal tissues and erodes the maternal blood vessels so that the surface is bathed in oxygen-carrying nutritious fluid. In addition the trophoblastic cells produce human chorionic gonadotrophin to stimulate the production of progesterone and maintain pregnancy.

The inner cell mass develops a fluid-containing cavity from its upper surface – the amniotic sac – which eventually expands to line the inner surface of the trophoblastic cell layer. This layer ultimately forms the chorion of the foetal membranes, and protects the embryo and foetus until parturition (childbirth).

## Three cell-producing layers

By the sixteenth day the inner cell mass has two layers of cells (see p.79):

The ectoderm, underlying the amniotic cavity, and the endoderm, which starts to form another fluid-containing cavity – the yolk sac. At about the same time the two layers start to develop a stalk, the beginnings of the umbilical cord, separating them from the trophoblastic layer.

At first the embryo receives nutrition from the trophoblast and the yolk sac, but gradually the thick trophoblastic layer, outside the area of implantation, regresses, leaving the implanted part to form a thick disc of tissue – the placenta. The stalk to the embryo grows blood vessels and incorporates the yolk sac as it becomes surrounded by the amniotic membrane.

In the third week the two layers of the inner cell mass produce a third layer sandwiched between them – the mesoderm (*above*).

The endoderm eventually forms the intestines, liver, gall bladder and pancreas. It also forms the larynx and trachea from a pouch at the upper end, and finally the lungs develop from small buds.

The mesoderm is the precursor of all the body's muscles, bones and blood vessels. The ectoderm supplies the surface of the body – skin, hair and nails; as well as the lining of the nose, mouth and anus; and the nervous system.

From the third week the most rapid development of the embryo takes place. It is at this stage that the embryo starts to transform from a flat disc of cells surrounded by fluid into a recognizable shape. The heart begins to develop as the basic tube of the primitive blood vessel divides and twists. The nervous system develops from a straight groove appearing along the middle of the ectoderm. The two edges curl towards each other and join to form the neural tube – the primitive spinal cord – and, at one end, a bulge appears that ultimately changes into the brain.

The central channel remains, becoming the central canal of the spinal cord and ventricles of the brain, in which cerebrospinal fluid will flow.

## The next two months

By the end of the third week (see *right*) the mesodermal tissue is arranging itself into somites – a segmental arrangement that can be seen in humans in the ribs of the thoracic cage. Each somite of muscle is eventually supplied by its own blood vessels and nerves.

At four weeks the embryo is 3 millimetres long. It consists of 25 somites, a large bulge of the heart, and has small ear pits in the head. It is not until the fifth week that a pair of pigmented discs appear on the head. These are the first signs of the eyes. In front is the small depression of the primitive nose. The limb buds, arms first and then legs, begin to grow by the sixth week.

At six weeks the embryo is just recognizable as a "human being", with a pulsating heart, forty somites and miniature ear flaps. At this stage the embryo is nearly 1.5 centimetres long and the hands, fingers and feet are just recognizable.

Next the somites lose their well-defined divisions and join up to form a solid mass of body tissue with recognizable thorax, abdomen and the beginnings of a neck. By the seventh week the bulbous protuberance is clearly a head with well-defined nose, ears, eyes and mouth. The pelvis is starting to form and the end somites lose their tail-like appearance as the sacral bone is developed.

It is during this rapidly changing period, from the fourth to the seventh week, that congenital malformations may occur if the tissues fail to fold or divide correctly. External influences, such as disease or drugs, may damage the embryo.

By the eighth week the embryo is approximately 2.5 centimetres long and recognizably human. It has distinct eyelids and separate fingers and toes developing.

From the end of the second month onwards development is mainly by growth and some alteration in proportion, so by the end of twelve weeks the embryo is 5 centimetres long and the mother's uterus is about the size of an orange. It is at this stage of development that the embryo is called a foetus.

## Growth of embryo

Actual size

Three-week embryo     Four-week embryo     Five-week embryo

Six-week embryo     Seven-week embryo     Eight-week foetus

## Maintaining pregnancy

Oestrogen
Progesterone
Corpus luteum
Human chorionic gonadotrophin
Implanted blastocyst
Foetus
Placenta

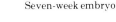

## Hormonal control of pregnancy

At the moment of conception the corpus luteum is producing progesterone and oestrogen to prepare the endometrium (the uterine lining) for implantation (*above*). Before the corpus luteum starts to degenerate, the trophoblast layer of the blastocyst begins to secrete human chorionic gonadotrophin. This starts about the fifth day and rapidly increases after implantation (1) to maintain the luteal secretions (2). This maintains the condition of the endometrium and suppresses further ovulation.

By the twelfth week the placenta produces more human chorionic gonadotrophin and also its own hormones (3) to maintain pregnancy. Until this stage the corpus luteum has grown in size, but it now begins to degenerate. If the increase in placental hormones does not keep pace with the reduction of those from the corpus luteum the placenta may detach and abortion occur.

# Pregnancy (Months 5 – 9)

■ **Mesoderm** Middle layer of the inner cell mass that ultimately forms the muscles, bone and cardiovascular system.

■ **Mitosis** Division of a cell to produce an identical one containing the same number of chromosomes.

■ **Neural tube** Precursor of the central nervous system which is formed by the ectoderm.

■ **Oestrogen** Ovarian hormone that helps to prepare the endometrium for implantation of the fertilized ovum.

■ **Ossification centre** Area in tissue from which bone is formed.

■ **Ovulation** Moment when an ovum escapes from the ovary.

■ **Oxytocin** Hormone produced by the posterior lobe of the pituitary gland that, with oestrogen, helps to initiate and maintain uterine contractions in labour.

■ **Parturition** Process of giving birth to a child.

■ **Placenta** Disc-like organ, about 25 cm in diameter, 3 cm thick and weighing about 600 g, composed of about 50 cotyledons and connected to the foetus by the umbilical cord. The placental function is to exchange foetal carbon dioxide and waste metabolites for oxygen and nutrition. It is able to excrete metabolites. It also secretes human chorionic gonado-trophin in the first three months of pregnancy and progesterone and oestrogen.

■ **Presenting part** Part of the foetus in the uterine cavity overlying the cervix.

■ **Progesterone** Hormone produced by the ovary in the second part of the menstrual cycle to maintain the endometrium during implantation of the blastocyst.

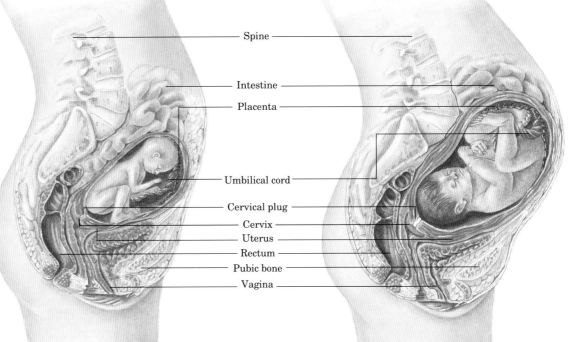

**Five-month foetus**   **Nine-month foetus**

Spine
Intestine
Placenta
Umbilical cord
Cervical plug
Cervix
Uterus
Rectum
Pubic bone
Vagina

## The continuation of pregnancy

At the end of the fourth month the foetus weighs about 125 grammes and is 12.5 centimetres long, the genitalia are developing and ossification centres appear. The foetus is moving in the amniotic fluid, and in the fifth month (*above left*) the mother may feel these movements – known as "quickening" – for the first time.

In the sixth month the foetus starts to produce a layer of fat under the skin and a fine hair – lanugo – covers the head and body, which is also covered by a greasy substance – vernix – that protects and lubricates the foetal skin, particularly during parturition. As the foetus grows the proportion of amniotic fluid decreases. At 28 weeks the foetus weighs about 500 grammes

and the heart-beat of 140 a minute can easily be heard; the foetus is now viable. Four to six weeks later the head engages in the pelvis and remains there until parturition. During the last two months (*above right*) it matures steadily and gains about 25 grammes weight a day to reach about 3 kilogrammes at birth.

### Foetal sex

Foetal sex depends on the male Y chromosome and foetal secretion of small amounts of testosterone at about the fourteenth week of pregnancy. About 106 boys are born for every 100 girls. However, there are fewer adult males owing to a greater male infant mortality.

### Twins

Twin births (*left*) occur about once in every 85 pregnancies. Identical twins (**A**) are always the same sex, developing from mitotic division of the same fertilized ovum. Each foetus has its own amniotic sac, but shares a single placenta. Fraternal, non-identical twins (**B**), develop from two separate ova that have been fertilized at the same time by different spermatozoa. They have separate placentae.

**Twin births**

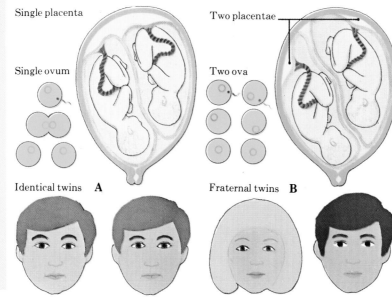

Single placenta

Single ovum

Identical twins   **A**

Two placentae

Two ova

Fraternal twins   **B**

## Circulation in the foetus and at birth

Throughout pregnancy oxygen reaches the foetus (*right*) from the placenta via the umbilical vein. Most of the blood passes through the liver. Some passes into the ductus venosus (3) and into the general circulation, after it has passed through the liver, to reach the heart. By this time it is only partially oxygenated.

In the heart much of the blood passes through the foramen ovale (1) into the left atrium and back into the systemic circulation. Most of the blood from the right ventricle is pumped through the pulmonary artery and a special bypass artery, the ductus arteriosus (2), to the systemic vessels. Only a small amount continues through the pulmonary system. Blood is then pumped through the systemic system and the two umbilical arteries to reach the placenta.

At birth (*far right*) the placental circulation ceases, the lungs expand, causing the ductus arteriosus and foramen ovale to close, and the circulation becomes independent.

## The mother in pregnancy

In the first trimester the hormonal changes may cause morning nausea and vomiting with fatigue, but there is little weight gain. After this period these unpleasant symptoms disappear and there is a steady gain in weight. This is only partly due to the growing foetus. The uterus, amniotic fluid and placenta contribute a certain amount. Over half the weight gain is a combination of fluid retention, which also softens joints and ligaments, and reserves of fats for the energy required during lactation. The mother will notice the "quickening" soon after the sixteenth week and then increasing awareness of movement and kicking until the end of pregnancy. There is often increased comfort – "lightening" – about the thirty-sixth week, when the head engages. This is due to a lowering of the uterus from the xiphisternum.

The mother's diet requires increased proteins, calories, vitamins, iron and calcium. Regular exercise and rest contribute to good health.

## Circulation before birth

- Head and arms
- Heart
- Lungs
- Liver

## The placenta

The placenta is fully developed by mid-pregnancy with 50 to 60 cotyledons each one supplied by a branch of the umbilical artery. It reaches an ultimate diameter of about 25 centimetres, with a surface area of some 10 square metres. The umbilical cord of two arteries, one vein and Wharton's jelly is usually attached to the centre. The function of the placenta is to exchange the foetal carbon dioxide and waste metabolites for oxygen and nutrition.

The placenta is bathed in maternal blood brought by spinal arteries through the uterine wall. The two circulations are separate, but leaks of foetal cells into the maternal blood may cause problems in Rhesus blood group incompatibility.

## Circulation after birth

- Placenta
- Body
- � Oxygenated blood
- Deoxygenated blood
- Mixed blood containing oxygen and waste materials

## The placenta – maternal and foetal blood vessels

Photo: Ralph Hutchings; courtesy Royal College of Surgeons

- ■**Prolactin** Hormone produced by the anterior pituitary lobe to initiate and maintain lactation.
- ■**Puerperium** Period of recovery from parturition with return to normal function.
- ■**"Quickening"** Moment at which the mother first notices foetal movements, between the sixteenth and eighteenth weeks.
- ■**"Show"** Loss of blood and mucus when the cervix starts to dilate at the onset of labour.
- ■**Somites** Segmental arrangement of the embryo in its early development.
- ■**Symphysis pubis** Joint between the two pubic bones in the front of the pelvis.
- ■**Trimester** Three months.
- ■**Trophoblast** Layer of cells surrounding the blastocyst.
- ■**Umbilical cord** Consists of two umbilical arteries, one vein and Wharton's jelly surrounded by the amniotic sac and extending from the centre of the placenta to the umbilicus It is of variable length usually 20 cm long.
- ■**Vernix** Greasy substance that covers the skin of the foetus.
- ■**Wharton's jelly** Soft connective tissue of the umbilical cord.
- ■**Xiphisternum** Inferior end of the sternum.
- ■**Yolk sac** Small sac produced by the endoderm layer of the inner cell mass. It initially helps in the embryo's nutrition.

# Childbirth

## Common diseases

■ **Abortion** Loss of the foetus before the twenty-eighth week of pregnancy. Criminal – termination that is illegally produced. Incomplete – when only part of the foetus or placenta has been expelled. Inevitable – pain and bleeding indicate that an abortion will take place. Missed – death of the foetus or embryo without expulsion. This can lead to formation of a carneous mole. Septic – abortion in the presence of uterine infection. Therapeutic – termination of pregnancy for legally acceptable reasons. Threatened – bleeding without pain.

■ **Antepartum haemorrhage** Accidental – bleeding from a normally situated placenta. Extra-uterine – bleeding from outside the uterus, e.g. cervix. Unavoidable – bleeding from a placenta praevia.

■ **Chorion-epithelioma** Malignancy occurring in a hydatidiform mole.

■ **Chromosome abnormalities** Many such abnormalities may abort, but some, e.g Down's syndrome, may survive.

■ **Ectopic pregnancy** Implantation in some place other than the uterus, such as the Fallopian tube or ovary. This may occur if there has been a delay in the movement of the fertilized ovum towards the uterus.

■ **Glycosuria** This may occur owing to diabetes mellitus appearing under the stress of pregnancy or to a change in the kidneys' ability to reabsorb glucose.

■ **Haemorrhoids** Can occur during pregnancy due to venous congestion below the presenting part and swelling of rectal and pelvic tissues.

**Sequence of normal birth**

A

B

C

D

E

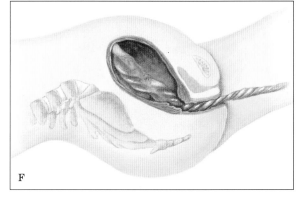

F

## Childbirth

At about 280 days after conception the foetus is expelled from the uterus (*above*). If pregnancy continues for too long the placenta begins to degenerate and the foetus is slowly starved of oxygen and nutrition.

Throughout pregnancy the uterine muscle has been contracting and relaxing. During the last few weeks these contractions have become stronger and are sometimes noticed by the mother. They help to keep the foetal head engaged in the pelvis, (**A**). Labour is divided into three stages: the first is from the onset of labour, with regular rhythmical uterine contractions, to the moment when the cervix is fully dilated. This may last from two to twenty-four hours or sometimes longer. The second stage is from the full dilation of the cervix to the moment of delivery and lasts less than two hours.

The third stage is the expulsion of the placenta, cord and membranes. The placenta has to be carefully examined to make sure that parts of it have not been left behind which could cause a haemorrhage.

## The first stage

The first stage starts with backache and regular uterine contractions that become gradually stronger and by the end occur about every two minutes, each lasting thirty to forty seconds. As the head is forced deeper into the pelvis it turns sideways (**B**) and the plug of cervical mucus is dislodged causing a "show" of blood. The cervix is gradually stretched, thinned and dilated until it is wide enough (**C**) for the head to pass into the vagina. If it has not already done so, the amniotic sac ruptures.

## Breech birth

## The second and third stages

The second stage starts with the cervix fully dilated (**C**). The head moves deeper into the pelvis and again turns so that the occiput is behind the symphysis pubis. The mother has a "bearing down" sensation and can help the powerful uterine contractions to force the foetus through the vagina. The vulva distend around the head and finally it is "crowned" when the greatest circumference is reached. The head extends (**D**) round the symphysis as it is pushed out. The baby's first breath is taken and is followed by a cry. This expands the lungs and starts respiration, as well as obliterating the foetal circulation.

The baby can now be delivered (**E**) with the next contraction as the smaller shoulders and body easily follow the large head. At this moment the mother is usually given an injection of oxytocin and ergometrine to stimulate uterine contraction and the separation of the placenta.

The third stage follows ten to twenty minutes later when the placenta has separated from the uterus and is expelled (**F**) along with the umbilical cord and membranes.

## A breech baby

A breech presentation (*above*) is when the buttocks engage in the pelvis instead of the head and delivery is complicated as the largest part, namely the head, is delivered last. Breathing cannot start until the same time as the placenta begins to separate with the obvious hazards of oxygen deprivation.

## The puerperium

After parturition the uterus returns to normal size in about a month with loss of lochia, which is initially fresh blood later becoming creamy in colour. Menstruation will return in about two months unless suppressed by prolactin secretion during lactation. Excess body fluid and fat gradually disappear and softened ligaments regain their strength.

The baby is born with excess fluid, which it loses in the first few days, and a raised haemoglobin count level, which falls slowly with the appearance of slight jaundice. This is the result of raised levels of serum bilirubin from haemolysis and the immature liver's inability to excrete it.

## Lactation

Throughout pregnancy the breasts have been enlarging under the stimulus of the anterior pituitary hormone, prolactin, to produce a watery milk (*below*). In the last few weeks before parturition colostrum is produced and secreted for the first few days. Then normal milk (**1**) is produced and the infant's sucking (**2**) stimulates the hypothalamus. This continues the production of prolactin and also oxytocin from the posterior pituitary gland (**3**), which causes the breast alveoli to contract and force – "let down" – the milk into the ducts (**4**).

## Hormones in labour

- Progesterone
- Oxytocin

## Hormones in labour

The story of hormones in labour (*above*) is not fully understood. In some way their interreaction adjusts the actual onset of labour. Throughout pregnancy the progesterone has made the uterine muscle (**A**) relaxed and unresponsive to the posterior pituitary hormone – oxytocin. Towards the end of pregnancy the progesterone production falls and there seems to be an increase in oestrogen. The combination of oxytocin and oestrogen initiates uterine contractions. The dilatation of the cervix (**B**) increases oxytocin output and further reduces progesterone, so the contractions become stronger, longer and more frequent (**C**). In the third stage oxytocin helps to obliterate the spiral arteries in the uterine wall and initiate lactation. Oxytocin can be synthesized and is often used to start contraction of the uterus if labour is delayed.

■ **Hydatidiform mole** Complication of an abortion or pregnancy due to continued growth of the chorion.

■ **Jaundice** Yellow discoloration that occurs when the bilirubin in the circulation rises above a certain level.

■ **Miscarriage** *See* Abortion.

■ **Multiple pregnancy** Homozygous – Identical cells, owing to mitotic division of the fertilized ovum forming separate units. Heterozygous – Fertilization of two or more ova at once.

■ **Placenta** Praevia – placenta is situated in front of the presenting part and may cover the internal opening of the cervix. Retained – complication of the third stage of labour in which the placenta does not separate normally from the uterus.

■ **Varicose veins** Distension of the veins in the legs due to pressure of the presenting part on the venous return in the pelvis.

■ **Vomiting** Early-morning vomiting is a common feature in the first trimester of pregnancy.

## Lactation

Colostrum

Milk

**Cross-references**

Cell 19
Heart 30-1
Blood groups 36
Endocrine 54
Skin 71
Reproduction 78–9
Nutrition 90–3

# Newborn to Puberty

## Definitions

### Infancy

- **"Grasp" reflex** Instinctive holding of an object in a baby's palm.

- **"Rooting" reflex** Instinctive turning of a baby's head towards something that touches its cheek.

- **"Step" reflex** Instinctive raising of a baby's foot when both are on a surface.

- **"Suck" reflex** Instinctive sucking when something is placed in the mouth.

### Teeth

- **Canine** One of four teeth with a single point projecting above the level of the adjacent teeth.

- **Cementum** Adherent connective tissue holding the ligament of the tooth to the jaw bone.

- **Crown** Enamel-coated portion of the tooth above the level of the gum.

- **Deciduous** "Milk" or temporary teeth

- **Dentine** Hard bone-like structure that forms the substance of the tooth.

- **Dentition** Number and arrangement of teeth.

- **Enamel** Coating of the crown of a tooth

- **Incisor** Four cutting teeth in the front of each jaw.

- **Molar** Three permanent grinding teeth on each side of both jaws.

- **Neck** Area of the tooth between the crown and root.

- **Premolar** Two deciduous and permanent grinding teeth on each side of both jaws.

- **Pulp** Soft centre of a tooth containing blood vessels, lymphatics and nerves.

- **Root** One or more extensions of the tooth into the bone.

**Locomotion and motor control**

## Infancy

The new-born baby lies with its knees drawn up. It reflexly grasps any object that touches the palm and when held upright automatically steps as the feet touch something. It roots and sucks the nipple automatically. These reflexes disappear with-in a few weeks. At a month the legs are straighter and by six weeks the head is lifted. The baby sleeps more often than not, but gradually the eyes move to focus on objects and at about six weeks smiling begins. By six months the birth weight is doubled and the child can sit unaided. At eight months the preliminary gur-glings of speech are heard and the thumb can be used. At about ten months crawling starts; the birth weight is trebled. The first step may be taken at a year old. The first words may be spoken during the next two or three months.

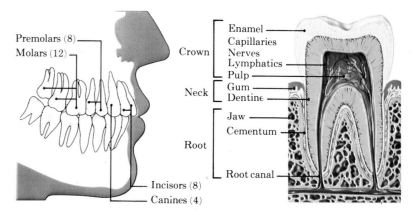

**Milk teeth (20)**

**Permanent teeth (32)**

Premolars (8)
Molars (12)

Crown

Neck

Root

Incisors (8)
Canines (4)

**Structure of a tooth**

Enamel
Capillaries
Nerves
Lymphatics
Pulp
Gum
Dentine

Jaw
Cementum

Root canal

## Teeth

Deciduous teeth start to appear at about six months; by the age of three the child has 20 teeth (*above left*) – 8 incisors, 4 canines, 8 premolars.

Between the ages of six and twelve years the deciduous teeth are shed and replaced by the permanent den-tition (*above centre*). A further six teeth, the molars, will appear in each jaw, and by the age of 25, a total of 32 teeth will be present.

The tooth consists of three parts (*above right*) – the crown, neck and root. The crown consists of dense mineral – enamel – surrounding the hard dentine, which has a soft centre – the pulp; the pulp is filled with blood vessels, lymphatics and the nerve, which reach it through the root canal. The neck adheres to the gum, and the root penetrates the bone, where it is held in place by a liga-ment and cementum.

# Stages of development from birth to puberty

| | Body | Senses | Mind | Speech | Social | Emotional | Moral |
|---|---|---|---|---|---|---|---|
| **At birth** | Almost immobile, but with many primitive reflexes e.g. sucking, grasping and rooting | Even at 3 to 4 days old a baby is able to follow sound from side to side and can tell a buzzer from a bell. | Rapidly learns to turn head to left or right to find food in response to a buzzer or a bell. | Crying or gurgling are the only sounds. | Can be divided into active, moderately active and quiet babies. | Feelings are almost undifferentiated. Babies are awake or asleep, active or inactive. | Newborn is amoral, having no values, no attitudes and no beliefs for the first 6 or 7 months. |
| **Up to 2 months** | Can focus eyes and coordinate stare. Can begin to lift chin when lying on stomach. | Loud noises may be painful. Baby smiles at a picture of a simplified face made up of dots and angles. | Forgets existence of object if blocked for more than 15 seconds. | The first 3 months is usually the period of maxiumus crying. | Has often established a distinct pattern of crying and sociability. Starts smiling at his mother at 4 to 6 weeks. | Beginnings of the differentiation of feelings into positive and negative. | |
| **2 – 4 months** | Able to raise head when lying on stomach and can turn it towards direction of speaker. | Shows signs of recognizing things. Can now appreciate 3-D. | Can tell mother from others but no idea that he has one mother only. Optical illusion of 3 delights him. | At beginning of period cooing starts. Responds to human voice and smiles when talked to. Sometimes chuckles. | Plays with objects (e.g. rattles) if placed in his hand. Beginnings of differential response to people around him. | At beginning of period first clearly defined signs of pleasure (smiling, chuckling, etc.) and negative emotions. | |
| **4 – 6 months** | Opens hand for contact with an object and grasps things using palm and fingers but not the thumb. | Brings things to mouth in order to explore them. Smiles at moving picture of a face. | Child now distressed when shown optical illusion of 3 mothers. | Cooing becomes more vowel-like and consonants also appear. | Beginnings of real differentiation between adults at end of period. | Pleasure responses to others become selective. Fear and anger begin to emerge separately. Signs of self-satisfaction. | |
| **6 – 8 months** | Can sit up without support for a short time, and stand with help. Can grasp using forefinger and thumb. | Attention begins to focus on expression rather than features of a face drawing. | Realization that hidden or out-of-sight objects continue to exist even when not seen. | Cooing begins to turn into babbling. Most common sounds are ma, mu, dar and di. | Beginnings of serious play but rarely with other children. | Fear of the unknown becomes developed. Attachment to known adults, and relationships becomes focused. | |
| **8 – 10 months** | Can crawl by the end of the period and pull himself upright to stand. Can hold feeding bottle on his own. | Depth perception is further developed. Child will avoid a visible drop from a surface. | Greater understanding of "object permanence". Child will search for objects in many different locations. | Repetitions of heard sounds become frequent but the results are often inaccurate. | Prefers to play with other people rather than by himself. Starts to imitate others. | Shows timidity towards strangers who are too familiar. Smiles at own image in mirror. | May show signs of withdrawal when admonished for bad conduct. |
| **10 – 12 months** | Can walk with help. Begins to stop putting objects in mouth. Bowel action starts to become more regular. | No longer likely to be upset by sudden, unexpected disappearance of an object. | With his first words begins to create and use symbols towards end of period. | First words are often nouns which serve multiple uses, i.e. "dog" may be used to describe any four-legged animal. | May show dramatic decrease in excitability and in attention and response when mother is present. | Differences between boys and girls seen in way they assert themselves and in readiness to touch others. | Understanding of parents' commands remains fairly primitive and tied to immediate situations. |
| **12 – 18 months** | Unaided walking at about 13 month. Helps to dress himself. Feeds himself. Creeps backwards down stairs. | Rapid increase in ability to use characteristic features of objects for identification purposes. | Up to 7 years increasingly links image of object with thought about it. | Repertoire of 3 – 50 words. Speech is largely telegraphic, e.g. "all-gone-food". | At end of period fighting and jealousy over toys emerges. Pleased with particular events or people. | Fear of strange objects, people and noises declines. At end of period shows jealousy over newly arrived babies. | Behaviour up to 6 or 7 years is shaped by threat of punishment of satisfaction of needs. |
| **18 months -2 years** | Runs but falls often. Climbs stairs holding rail. Girls reach half adult height after boys, just before 2 years old. | Can recognize displacement even though he has not seen object moved. | Thinks everyone has same view of object as he has, i.e. if covers eyes and cannot see you, you cannot see him. | Can string two words together (noun and qualifier). Grammatical use of articles, plurals, etc. | Solitary play declines. Parallel play with little contact between children in one room becomes more common. | At end of period fear of imaginary creatures and dark begins. More sensitive to ridicule. | Child's comprehension of parents' moral code consists of simple good/bad distinction. |
| **2 – 3 years** | Control over sphincter muscles allows toilet training. Bed wetting increasingly rare. Self-feeding improved. | Shows ability to distinguish between primary body parts and features; developing skill at drawing them. | Progress in ability to classify many sorts of objects according to one or more shared characteristics. | Over period able to put 3 or 4 words together. Structuring of language begins, use of tenses, plurals, order, etc. | By 30 months will be helping with domestic chores. Co-operation between children playing together increases. | Boys express more physical assertiveness and anger than girls. Attempts by both sexes at more independence. | Child likely to show substantial guilt-like reactions, though still situation-specific. |
| **3 – 5 years** | At end of period girls are slightly ahead of boys in skeletal development. The brain is 75% of its adult weight. | Often confuses the letters b and d, p and q. Ignores straightness, so confuses D and O. | Realizes apparent change (e.g. water/ice) is not real change, but is unlikely to understand why this is so. | Sentences become longer and more complex. Increasing grasp of basic grammar, like active passive. | Peers start to influence child's behaviour. Identification with same-sex parent is at its strongest. | Increased satisfaction at accomplishing a self-set task. Rising sensitivity to feelings and responses of others. | Still highly egocentric but beginning to show marked guilt and self-regulation. |
| **5 – 7 years** | At 6 years old child's brain is 90% of its mature adult weight. Nerve fibres, etc., are almost complete. | Focuses on significant features of object or event. Increasingly able to discriminate between letters of the alphabet. | Child can use concepts and rules, but deals almost exclusively, in the beginning at least, with here and now. | Masters irregular endings (i.e. mice rather than mouses). Vocabulary becomes more complex and school related. | Turns to peers rather than adults. Play teaches social roles and individual limits. | Child actively engaged in testing self-image. | Child's judgement of "right" changes. Respect for, rather than fear of, law. |
| **7 – 11 years** | Rise in strength, speed and co-ordination. Skills now include those that need instruction and practice, e.g. skiing. | | Can now conserve and see parts of the whole. Is able to arrange objects according to size, weight, etc. | By end of period has almost complete mastery over complex grammatical rules. | Social interactions strongest with members of the same sex; minimal social interaction between sexes. | Becomes less physical and more verbal in relationships with peers. | Child does things either to win approval or because the law says so. |
| **11 – 13 years** | Girls put on sudden growth spurt and by end of period they average 2 1/2 years ahead of boys. Menstruation begins. | Increasingly is able to perceive non-obvious relationships among apparently unrelated events and objects. | Able to think around a subject, testing many solutions in his mind. Use of abstract rules to solve problems. | Ever widening vocabulary of more technical and specialized words. Greater range of terms to describe phenomena. | The sharing of possessions, feelings and plans is still primarily with same-sexed peers. | Alternating periods of withdrawal and gregariousness. Sexual feelings start to emerge. | Actions are guided according to universal ethical principles. |

# Puberty to Old Age

## Puberty

Puberty is the moment when sexual maturation has reached the point where sexual reproduction is possible. A period of development around puberty takes about two years and will usually start earlier in girls than it does in boys.

In girls, between 10 and 16, there are changes in the subcutaneous fat – hips and shoulders become more rounded, the breasts develop and the pubic and axillary hair grows. The first menstruation – menarche – is a dramatic event establishing a definite point of maturity.

In a boy, between 12 and 17, the shoulders broaden, the muscles strengthen and the genitalia develop and darken to become covered with pubic hair. The larynx lengthens and the voice "breaks" to become deeper. Spontaneous erections occur; the nocturnal emissions of sperm are less dramatically the sign of sexual maturity than the menarche.

## Adolescence

Adolescence is the time of greatest physical and emotional change, which can be disturbing not only to the individual concerned but also to the family. It is a time when social, intellectual and sexual interests expand and broaden. The young person, no longer a child and already more or less physiologically mature, must gradually loosen the ties associated with the home and parents. This involves a questioning of parental wisdom and morality, which is often replaced by less desirable leadership and a tendency to truancy and even crime.

The intense emotions of adolescence are often released as an obsessive or hysterical hero worship, particularly of such heroes as sportsmen and pop stars. Screaming at pop concerts, for example, is one way the adolescent can release pent-up emotions. Such periods of great enthusiasm may be followd by ones of almost depressive lethargy.

The stresses of adolescence and the feeling of independence may lead to experimentation with drugs as an escape from the conflicts of a demanding adult world. This turbulence may lead to depression.

## The young adult

As the physical and psychological turmoil of adolescence settles young adults are at their prime. There is a better relationship with family and society and a greater awareness and control of emotions and sexual desire. The first tentative adolescent meetings with the opposite sex have passed and mature relationships develop. Work may give a felling of achievement and independence – a confidence previously lacking.

## Middle age

Usually the most stable period of life starts around the age of 30. It is usually a time of relative financial and emotional security built around the family and growing children, in those who have married, or friends and work in those who remain single. Sexuality is less urgent but often more satisfactory.

In the woman the menopause – the final cessation of menstruation – may be followed by a time of hormonal disturbance leading to hot flushes, tensions and sometimes depression. If this coincides with the storm of her children's adolescence and her husband's demands at work, she may feel her life is a failure; in fact, it is a time of achievement.

Although physical health usually remains good there is a tendency to gain weight, accompanied by a slight, but definite loss of athletic skills. In the later years there is an increase in the death rate, particularly from heart disease and cancer of the lungs.

## Old age

In Western society old age seems to start arbitrarily on retirement. This is a psychological shock, as many still feel fit and active, and find difficulty in accustoming themselves to days of idleness and boredom, often with increasing financial anxiety. The stiffness of early arthritis, the loss of visual acuity and minimal deafness may produce a slower reaction time, so making falls more common.

The adrenal glands continue to secrete sex hormones and the slight androgenic effect may produce slight baldness in women, while in men there may be the oestrogen effect of female fat distribution and less beard growth. Cancer and other illnesses are common and often fatal.

## Senescence

This is the period of real ageing and may begin seriously before or long after the arbitrary point of retirement. The skin loses its elasticity and gains wrinkles; the hair loses its colour, becoming white as it fills with microscopic bubbles of air; the senses of vision, hearing and taste deteriorate; the joints become rough and arthritic, while the muscles weaken. The bones in the skeleton soften as the structure deteriorates and calcium is removed, causing the bent spine of the elderly.

The nervous system is involved with loss of nerve cells, leading to failure of memory, first for recent events and later a gradual dementia may occur. The cerebellum is affected causing a loss of co-ordination, tremor of the hands and hesitation in speech. The heat centre in the thalamus is less responsive and a cold environment may cause a drop in body temperature – hypothermia – that can lead to coma and death. Deterioration in the heart and lungs may produce irregularities of cardiac rhythm and often breathlessness.

Dietary problems are not only financial but also practical, involving preparation of the correct kinds of food when arthritic, absent-minded and breathless. Chewing is difficult and failure to eat roughage leads to constipation and abdominal pain.

In some people these ageing processes take place faster in one system than another. This can be seen with dementia occurring while the body is fit or osteo-arthritis crippling an otherwise healthy woman.

Illness and disease may bring death before the possibility of social problems, such as the paranoid, incontinent and arthritic grandparent who fails to eat properly and refuses help. These social, economic and family pressures of senescence are enormous and need help from visiting nurses, social workers and the family doctor unless admission to hospital is the only way of giving sufficient basic nursing care.

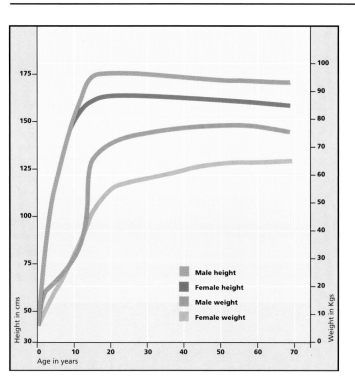

**Weight and height** The graph shows a steady increase in males and females until the early teens, when there is a short growth spurt reaching a maximum. The height loss with ageing is caused by loss of spinal bone density.

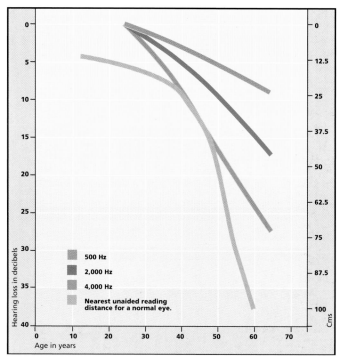

**Hearing and vision** As one ages, the nearest point on which the unaided eye can focus dramatically recedes around the age of 40. The ability to hear also declines with the highest pitches deteriorating faster than lower frequencies.

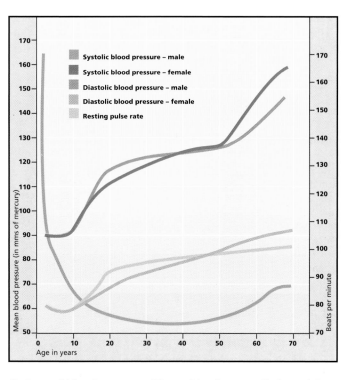

**Pulse and blood pressure** The rapid pulse rate of infancy falls to a low rate during adulthood and is accompanbi4ed by a rise in blood preeusre that is greater in systolic than diastolic levels, caused by arterial hardening.

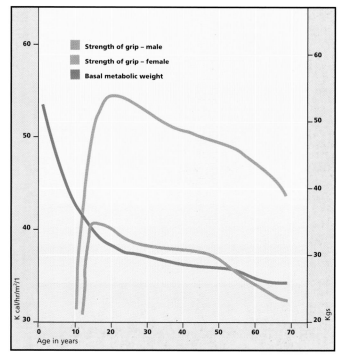

**Physical strength and basal metabolic rate** The development of greater strength in males is followed by steady deterioration with age. The metabolic rate declines rapidly at first and then slows down in life as cell metabolism declines.

# Metabolism and Nutrition

## Food requirements

Eating a variety of foods gives a good balance of nutrients. Nutrients are essential for the growth, maintenance and repair of the living body.

The main nutrient groups are carbohydrate, protein, fat, vitamins and minerals. Water is also essential for life, but is not usually considered a food or nutrient. Usually foods are complex mixtures consisting of a variety of carbohydrates, fats, protein and water. Vitamins and minerals are found in smaller amounts.

## Carbohydrates

Carbohydrate is the major source of energy in the diet. There are two main kinds of carbohydrate – sugars and starches. Sugars are found in jam, confectionery, table sugar, honey, fruit and fruit juice. Starches are found in potatoes, bread, rice and pasta. Another important source of carbohydrate is known as non-starch polysaccharide (NSP), such as cellulose, pectin, beta glucan and guar gum. NSP is found in fruits, vegetables, beans and cereals. These non-starch polysaccharides, together with lignin (which is not considered a carbohydrate), are referred to as Dietary Fibre. This mixture of indigestible materials is not absorbed into the body. Instead it adds bulk to the faeces due to its water building capacity and helps carry waste materials through the intestine.

Having been digested and absorbed in the blood stream as glucose, carbohydrate may be either utilized directly, temporarily stored in the liver as glycogen, or if there are excess amounts, converted into fat. These three processes are controlled by the hormones insulin and glucagon as well as those that are produced by both the adrenal and thyroid glands.

## Proteins

Protein is the nitrogen-containing, body-building material found in most tissue enzymes and muscles. It is broken down by digestion into amino acids required for the growth and repair of body tissues and it provides a source of energy. The liver can synthesize most amino acids by a process called transamination. However, there are eight amino acids which have to be obtained from dietary sources – these are known as the essential amino acids. If a protein contains all the essential amino acids it has a high biological value (HBV). Proteins with a relatively low concentration of one or more essential amino acids have a low biological value (LBV). Animal proteins usually contain HBV protein while plant foods supply mostly LBV proteins. By combining plant proteins the amino acid of one protein may compensate for the deficiencies of the other, resulting in a combined protein of a high biological value. This is known as the complementary action of proteins. An average estimated requirement for protein is 1g/kg body weight. Protein can be used for energy or changed into fat.

## Fats

Fat is a concentrated source of energy and it is metabolized to produce carbon dioxide and water. If too much is used, ascetic acids and ketones are produced.

Fat is a carrier for the fat-soluble vitamins A, D, E and K, and it also contains the essential fatty acids which are important in the formation of cell membranes particularly in nerve tissue. Fat is usually described as saturated or unsaturated according to the proportions of fatty acids present.

Generally, animal fats contain saturated fatty acids and vegetable fats contain unsaturated fatty acids. A high fat intake and in particular a high saturated fat intake has been associated with a raised blood cholesterol level which is one of the risk factors for coronary heart disease.

## Vitamins and minerals

Vitamins are complex organic substances that are needed in very small amounts for many of the processes carried out in the body. Most vitamins cannot be made by the body so have to be provided by the diet. The fat soluble vitamins are A, D, E and K; the water soluble vitamins are the vitamin B group and C.

Minerals are inorganic substances required by the body. They are needed for the formation of bones and teeth, for nerve function, for components of enzyme systems and as essential constituents of body fluids and tissues. The major minerals required are calcium, phosphorus, magnesium, sodium, potassium and chloride, and are needed in larger amounts than the trace elements which include iron, zinc, iodine, flouride, selenium and copper.

## Energy requirements

The energy requirements of individuals vary widely depending on their basal metabolic rate (BMR) and their level of activity. The BMR can be thought of as the rate at which a person uses energy to maintain the basic functions of the body and is always measured when a person is completely at rest.

The Basal Metabolic Rate is increased by five factors:

**Age** – infants and young children have a proportionately higher BMR than adults as energy is required for growth.

**Pregnancy** – Increased BMR in the mother to account for the baby's BMR.

**Sex** – Men usually have a higher BMR as they tend to have more muscle.

**Food** – Extra heat is produced after eating and more energy is therefore needed.

**Temperature** – Fever increases BMR by 10 per cent for each 1°C rise in temperature.

The Basal Metabolic Rate is decreased by four main factors – sleep, starvation, low thyroid activity and prolonged excessive chilling.

## Calories

Energy is measured in calories or kilojoules and is provided in the diet by carbohydrate, protein, fat and alcohol.

1 gram of carbohydrate provides 3.75 cal/16 KJ.

1 gram of protein provides 4 cal/17KJ.

1 gram of fat provides 9 cal/37 KJ.

1 gram of alcohol provides 7 cal/29 KJ.

Therefore, fat and alcohol provide almost twice as much energy than protein or carbohydrate.

## The Major Food Categories

| Carbohydrate | Protein | Fat | Vitamins and Minerals |
|---|---|---|---|
| **Sugars:** Jam, honey, marmalade, biscuits, cakes, confectionary, fruit juice.<br>**Starches:** Bread, potatoes, rice, pasta.<br>*Dietary fibre:* Wholegrain cereals, fruit, seeds, pulses, vegetables. | **High biological value:** Meat, fish, cheese, milk, yoghurt, eggs, chicken.<br>**Low biological value:** Peas, beans, lentils, nuts, cereals. | **Saturated:** Butter, hard fat on meat, cream, full fat cheeses, chocolate.<br>**Polyunsaturated:** Soft margarines, vegetable oils, oily fish. | *See* chart below. A varied diet provides enough for average daily adult requirement. |

## The Major Vitamins and Minerals

| Vitamins and minerals | Main function | Main sources | Nutrient Intake Male | Female | Deficiency |
|---|---|---|---|---|---|
| A (retinol) | Essential for vision at night; necessary for healthy skin and mucus membranes | Butter, fortified margarine, chesse, milk, egg yolk, liver, oily fish, dark green vegetables, tomatoes and carrots | 700μg/day | 600μg/day | Night blindness, lowered resistance to infection, skin infections |
| $B_1$ (thiamin) | Required for the continuous release of energy from carbohydrate; necessary for brain and nerve functions | Cereals, nuts, pulses, meat (esp. pork) and fortified cereals | 1.0mg/day | 0.8mg/day | Leads to beri-beri |
| $B_2$ (riboflavin) | Involved in the energy release from food especially fat and protein; required for healthy skin | Liver, milk, cheese, yoghurt, eggs, green vegetables, yeast extract and fortified cereals | 1.3mg/day | 1.1mg/day | Causes cheilosis (cracked and dry lips) |
| Niacin (nicotinic acid) | Involved in the release of energy | Liver, meat, fish, wholewheat flour, yeast, bran and instant coffee | 17mg/day | 13mg/day | Leads to pellagra (inflammation of the mouth, anaemia, skin rashes and diarrhoea) |
| $B_6$ (pyridoxine) | Essential in many of the amino acid reactions of the body. | Whole wheat cereals, fish, nuts and liver | 1.4μg/day | 1.2μg/day | Anaemia, dry and cracking skin |
| $B_{12}$ (cyanocobalamin) | Required for the production of red blood cells and nerve fibres | Liver, meat, eggs, cheese, milk, fish, and fortified breakfast cereals | 1.5μg/day | 1.5μg/day | Causes pernicious anaemia (degeneration of nerve cells) |
| Folic acid | Required for use with vitamin $B_{12}$ | Liver, yeast extract, and green leafy vegetables | 200μg/day | 200μg/day | Megaloblastic anaemia |
| Biotin | Involved in lipogenesis, gluconeogenesis, and catabolism of branched chain amino acids | Synthesised by intestinal flora. Rich sources include offal and egg yolk | 10-200μg/day | 10-200μg/day | Rare except in individuals who consume large amounts of raw egg white or undergo long periods of Total Parental Nutrition |
| Pantothenic acid | Part of coenzyme. A molecule which plays an essential role in the catabolism of all macro-nutrients to yield energy | Animal products, cereals and legumes | 3-7mg/day | 3.7mg/day | Experimental dietary deficiency causes fatigue, dizziness, muscle weakness and gastro-intestinal disturbances |
| C (ascorbic acid) | Necessary for growth and repair of tissue and aids iron absorption | Citrus fruits, green vegetables and potatoes | 40mg/day | 40mg/day | Results in bleeding gums and poor wound healing; scurvy follows |
| D | Promotes the absorption of calcium and phosphorus and is thus essential for bones and teeth | Sunlight, oily fish, cod liver oil, egg yolk, fortified margarines and breakfast cereals | | | Leads to rickets in children and osteomalacia (soft bones) in adults |
| E (tocopherol) | Acts as an anti-oxidant protecting cell membranes from damage by oxidation | Vegetable oils, cereals, nuts and eggs | Not known | | Rare but can occur in premature babies |
| K | Necessary for the normal clotting of blood | Dark green leafy vegetables, cereals and pulses | Not known | | Increased clotting time |
| Calcium | Healthy bones and teeth, blood clotting, nerve transmission and activity of some enzymes | Milk, cheese, yoghurt, canned fish, white flour, bread and green leafy vegetables | 700mg/day | 700mg/day | Leads to rickets in children and osteomalacia in adults; a deficiency can result from a lack of Vitamin D |
| Sodium | Regulation of the water balance in the body; essential for muscle and nerve activity | Salt – either added to foods in processing or at the table or during cooking | 1600mg/day | 1600mg/day | Fatigue, nausea and cramps; thirst is experienced |
| Potassium | Part of body fluids; nerve and muscle function | Oranges, bananas, vegetables, chocolate and instant coffee | 3500mg/day | 3500mg/day | Weakness, mental confusion and in severe cases heart failure |
| Magnesium | Involved in nerve and muscle functioning and some of the enzymes involved in energy utilization | Widespread especially of vegetable origin; wholegrain cereals, nuts and spinach | 300mg/day | 270mg/day | Depression, fits, tiredness and if extreme a heart attack |
| Phosphorus | Essential for healthy bones and teeth and is a major constituent of all cells; required for the formation of nuclei acids and ATP | Meat, fish, cheese, eggs, nuts, pulses, rice, bread and yeast | 550mg/day | 550mg/day | Dietary deficiency is unknown in humans |
| Zinc | Essential for growth and repair; involved with enzyme activity and taste perception | Milk, cheese, meat, eggs, and shellfish, wholegrain cereals and pulses | 9.5mg/day | 7.0mg/day | Rare, but can cause delayed puberty and retarded growth |
| Iodine | Production of thyroid hormones | Seafood, seaweed and iodized salt | 140μg/day | 140μg/day | Goitre, cretinism and myxoedema |
| Fluoride | Associated with the structure of bones and teeth | Fluoridated water, tea, fish with bones and toothpaste | Varies according to the area | | Contributes to tooth decay |
| Selenium | As an antioxidant it protects cell membranes against oxidation | Cereals, meat, fish, offal, cheese and eggs | 75μg/day | 60μg/day | Keshan disease (a type of heart disease) |
| Iron | Production of haemoglobin and myoglobin | Red meat and offal, bread, flour, eggs, vegetables and fortified cereals | 8.7mg/day | 14.8mg/day | Iron deficiency anaemia |
| Copper | Associated with a number of enzyme systems | Green vegetables, fish and liver | 1.2mg/day | 1.2mg/day | Very rarely may cause anaemia |

# Food for Health

## Diet and the whole body

The healthy human body is stimulated by the sight and smell, and even the thought, of food; and it is the brain that is responsible for co-ordinating these stimuli.

The limbic system and the hippocampus and cingulate gyrus help to activate the appetite centre in the hypothalamus. The olfactory lobes and visual cortex relay messages through the thalamus and mid-brain to initiate salivation and gastric secretion via the autonomic nervous system. At the same time a fall in the level of blood sugar together with the biological rhythms of the body will often coincide and increase the sensation of hunger.

These stimuli are reinforced by the action of eating, which activates the taste buds in the tongue and thus increases further salivation. The swallowing of food into the stomach sets in action gastric secretion and gastric hormonal release – gastrin – to maintain gastric and initiate duodenal secretions. This, in turn, produces further hormones to activate pancreatic and bile production.

In the graph (see *below*) five periods in the life of either a man or woman (and also the periods of pregnancy and lactation in a woman's life) have been selected. Each block represents an essential nutrient in a healthy diet. The intake of a nutrient can be compared as a percentage of the maximum ever required and the quantity by the thickness of the appropriate block.

The extraction of these nutritional elements occurs after the enzyme digestion. They are either used immediately in order to replace and repair worn-out and damaged tissue or alternatively transformed into energy and dissipated as heat. Some nutrients are stored in the liver or transformed into fat and transferred to the fat depots throughout the body. The carbon dioxide produced is exhaled by the lungs; the water is excreted by the kidneys and the amino-acids either broken down into urea, uric acid, creatinine and ammonia, or reformed into new protein by the liver. Cholesterol is excreted in the bile.

All this activity is dependent on small amounts of some of the essential nutrients to help maintain the intra-cellular enzymes in healthy metabolic order. The connective tissue starts to disintegrate without vitamin C – ascorbic acid; the haemoglobin fails to form without adequate supplies of iron and calcium absorption; and bone production is impaired without vitamin D.

The metabolism of every cell depends upon receiving the right substances via the liver and the body's circulation. Hormones are responsible for balancing this activity, many of them under the overall control of the anterior lobe of the pituitary gland.

As the diet is digested and assimilated into the healthy body the activity of the reticular system in the brain stem is reduced and a feeling of contentment and drowsiness begins to cloud the consciousness.

The heat produced by the metabolic energy of the digestive process and the specific dynamic action of the various foods also combines to add to the warm contentment of a satisfactory meal.

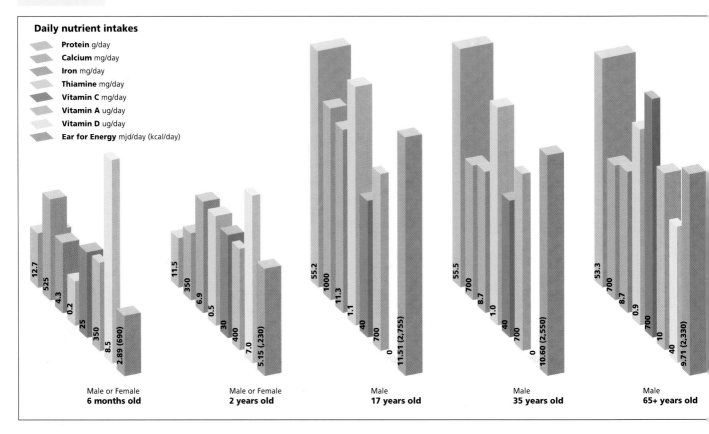

**Daily nutrient intakes**

- **Protein** g/day
- **Calcium** mg/day
- **Iron** mg/day
- **Thiamine** mg/day
- **Vitamin C** mg/day
- **Vitamin A** ug/day
- **Vitamin D** ug/day
- **Ear for Energy** mjd/day (kcal/day)

| Male or Female 6 months old | Male or Female 2 years old | Male 17 years old | Male 35 years old | Male 65+ years old |
|---|---|---|---|---|
| 12.7 | 11.5 | 55.2 | 55.5 | 53.3 |
| 525 | 350 | 1000 | 700 | 700 |
| 4.3 | 6.9 | 11.3 | 8.7 | 8.7 |
| 0.2 | 0.5 | 1.1 | 1.0 | 0.9 |
| 25 | 30 | 40 | 40 | 10 |
| 350 | 400 | 700 | 700 | 700 |
| 8.5 | 7.0 | 0 | 0 | 40 |
| 2.89 (690) | 5.15 (230) | 11.51 (2,755) | 10.60 (2,550) | 9.71 (2,330) |

## Energy requirements

■ **Specific dynamic action** Amount of energy required to release the potential energy in food. More energy is used in metabolism of protein than either fat or carbohydrate, which means that protein gives less useful energy to the body.

■ **Thalamus** Mass of grey matter at the base of the brain concerned with sensory nervous transmission.

■ **Urea** Simple nitrogen-containing substance that is a body metabolite.

■ **Visual cortex** Situated at the back of the cerebral hemisphere, it is the area of grey matter concerned with vision.

### Energy requirements

The body's need for energy from the diet varies not only with the individual's activity, whether male or female, disease and climate, but also with age. The size of the figures in the diagram (*above*) show the comparative energy requirements from birth to adulthood.

Up to approximately two years old, the energy requirements of a rapidly growing child are greater than anyone in proportion to size; however, as the diagram shows, the body's energy requirements fall with increasing age, until by old age, when the body's metabolism is at its slowest, the need is far less.

**Cross-references**

Cell 14-16
Muscles 28
Digestion 45
Endocrine 53
Nervous System 60–1

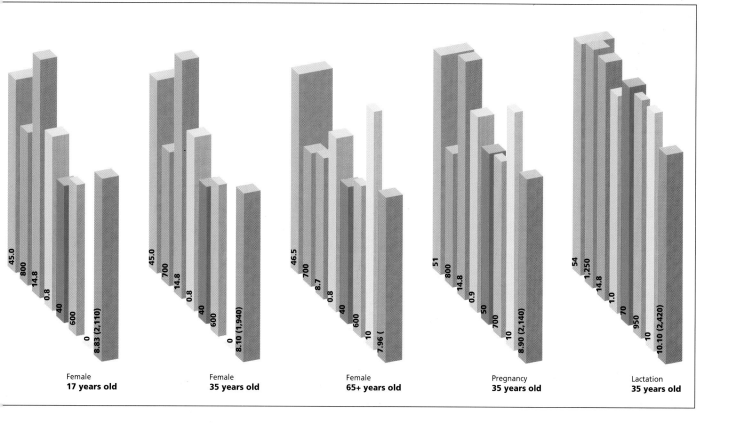

| Female 17 years old | Female 35 years old | Female 65+ years old | Pregnancy 35 years old | Lactation 35 years old |
|---|---|---|---|---|
| 45.0 | 45.0 | 46.5 | 51 | 54 |
| 800 | 700 | 700 | 800 | 1,250 |
| 14.8 | 14.8 | 8.7 | 14.8 | 14.8 |
| 0.8 | 0.8 | 0.8 | 0.9 | 1.0 |
| 40 | 40 | 40 | 50 | 70 |
| 600 | 600 | 600 | 700 | 950 |
| 0 | 0 | 10 | 10 | 10 |
| 8.83 (2,110) | 8.10 (1,940) | 7.96 | 8.90 (2,140) | 10.10 (2,420) |

# Index

# Index